Highway to My Way
A Trip of Fears, Jeers, Cheers, and Tears
By
Fred T. Jackson

I0190495

DeeJak's
PUBLISHING COMPANY

DeeJak's Publishing Company
Charlotte, North Carolina
www.deejakspublishing.com

DeeJak's Publishing Company
7209-J East W.T. Harris Blvd # 279
Charlotte, NC 28227-1004
www.deejakspublishing.com

Editorial: Micheal Furham
Cover and layout design by: Crystal Jeffrey
ISBN: 978-0-9857903-4-9
Library of Congress Control Number: 2013956232
Because of the dynamic nature of the Internet, any web addresses or links contained in this book may have changed since publication and may no longer be valid. The views expressed in this work are solely those of the author and do not necessarily reflect the views of the publisher, and the publisher here by disclaims any responsibility for them.

ACKNOWLEDGEMENTS

I wanted to thank the people who helped shape me into the person I am today.

The Most influential, I believe, were: Collin William Jackson (Eddie): Brother.

Benjamin E. Morgan: Agriculture Teacher
Collin Jackson: Father

Altheria Smith-Patton: English Teacher

DEDICATION

This book is dedicated to Black Jimmy, my family, friends and the many dedicated military personnel with whom I served. It is my effort to ensure that when my epitaph is written there can perhaps be more said than just the standard head stone remarks. Life has a way of going so fast. I like to share with you some memories of my life to help fill in some of the blanks caused by my frequent and long absents during my military service.

TABLE OF CONTENTS

PREFACE

After receiving my record of military service and thoroughly reviewing it, I felt compelled to write a historical memoir. As I began work on this project, I was faced with a huge dilemma: how much family history do I include and how do I relate it to who I am, while keeping it relevant and interesting.

To make my story inviting and interesting to family and non-family alike, I decided to make my military career the centerpiece and inject family history only where I felt it supported the objective of my memoir and improved the narrative.

This memoir, being so important to me, demanded that I find a way to make others feel some measure of empathy. I chose not to attempt to center on family history except where it was needed to help connect me with the broadest possible readership.

Though this memoir is meant mainly for family, I hope it can be enjoyed by everyone who might read it for what it is – a true narrative spanning decades and encompassing a military career as well as the epic civil rights struggle between justice and injustice being waged during the entire time. How I was affected by that struggle and the impact it had on family, friends and others is of significant relevance to this memoir. My perseverance in the face of tremendous adversity makes this an engaging memoir. I hope it will inspire others.

I have included many names of people, places and incidents to validate my story because they can be verified and cross-checked. Most of the incidents I've opted to include from my military service years can easily be verified. I know everyone has their own story, this is mine, and I hope folks reading it find it interesting and worth their time.

PROLOGUE

"How many roads must a man walk down before they call him a man?"

- Bob Dylan

After I learned to read at around the age of three and a half years old, I developed an interest in writing. Even as a child, I thought about writing my life story. I was too young to know you had to experience life in order to have a life story. My story has three distinct chapters, pre-military service years, military service years, and post-military times. The first two chapters are dealt with extensively in this memoir.

I lived the first eighteen years of my life within a twenty-five mile radius in rural Anson and Union counties in North Carolina. The next twenty-one years were spent roaming the world and sightseeing across five different continents while on various military assignments.

The rest of my life is still a work in progress. It is my intent to deal mainly with just the first two chapters of my life in this memoir.

By the time I reached age 73, it dawned on me that my life had, by no means, worked out the way I'd wanted, but it's been an engaging and exciting ride. Everything included in this memoir is based on actual experiences.

When I set out to write my life story, I realized I had, in addition to weak writing skills, two other problems to solve. The first problem was how to tell the story in a way to interest the broadest possible readership. Next, I wanted to find the best way to solve the first problem without losing the substance and context of the narrative, or sight of the objective of the memoir.

My objective is to accurately present the circumstances that propelled me from farm life in rural North Carolina into and through an interesting 21-year career in the United States Navy. Some of these episodes may be hard to believe, but they are all based on actual accounts. Here is my attempt to deal with chapters one and two of my life. Stay tuned for chapter three.

****

"No way of thinking or doing can be trusted as truthful without proof."

- Henry David Thoreau

PART ONE

Pre-Military Days:
A Time of Youthful Growth

****

****

CHAPTER 1 – FEARS

"Taking a new step, uttering a new word, is what people fear most"

- Fyodor Dostoyevsky

In 1958, at the age of 18, I joined the U.S. Navy and left home. As with Marco Polo, many years passed before I returned home. This is the recounting of that 21-year journey taken from my memory, military records and cruise books.

To give an overview of the extent of my travels, I will provide a list of some of the places I visited. Maybe I should title this memoir "Trying to Get Some Respect," "The Highway to My Way," or "From Dirt Lanes to Sea Lanes," or perhaps it could be called "What's Travel Got to Do with It?" Or maybe something like "A Trip of Lifetime Proportions" or perhaps simply "A Trip to Be Remembered." Hopefully, this list will grab your interest and provide some appreciation of my extensive travels.

The following is the opening excerpt from my speech "Tribute to the Teachers" given at the first class reunion held at Polkton High School.

" In the years since pomp and circumstance was played for my class of 1957, I've sipped Champaign on the French Rivera, and dined with African Kings. I've witnessed the devastation of war in South East Asia, and got caught in political riots in Spain. I even got myself captured by a lynch mob right here in this town. None of which gave me as much excitement as this opportunity to stand here today and try to pay tribute to our teachers here at Polkton School."

- Fred T. Jackson

- Great Lakes, Illinois – Philadelphia, Pennsylvania California

- Newport, Rhode Island – Orlando, FloridaPortsmouth, Virginia

- Jacksonville, Florida – Charleston, South Carolina – Lemoore, California

- Hawaii – Wake Island – Okinawa – New York – New Jersey

- Bomba, Libya – Pollensa Bay, Mallorca – Toulon, France – Mexico

- St. Florent, Corsica – Messina, Sicily – Taranto, Italy – Corfu, Greece

- Brindisi – Pilos, Greece – Stenosia, Greece – Athens, Greece

- Palermo, Sicily – Barcelona, Spain – Malta – Genoa, Italy – Mallorca

- Bethesda, Maryland – Camp Lejeune – Norfolk, Virginia – Tripoli

- Charleston, South Carolina – Grand Bahamas – Jacksonville, Florida

- Fort Lauderdale, Florida – Rota, Spain – Suez Canal – Bermuda

- Mombasa, Kenya – Alexandria, Egypt – Naples, Italy – Cairo, Egypt

- Cannes, France – Palma, Spain – Danang, Viet Nam – Saigon

- Hue, Viet Nam – Mekong Delta, Viet Nam – Quang Tre, Viet Nam

- Newport, Rhode Island – Patrick Air Base, Florida

- Adak, Alaska – Yokosuka, Japan – Reykjavik, Iceland – Quang Tre

- Orlando, Florida – Guantanamo Bay, Cuba – Port Au Prince, Haiti

- Nice, France – Gulfe Juan – Suda Bay, Crete – Sardinia

- Messina, Italy – Gibraltar – Valencia, Spain – Tripoli, Libya

- Manila, Philippines – Subic Bay – Partras, Greece – Piraeus, Greece

- Toulon, France – Monaco – Bomba, Libya

I visited many other destinations in the United States and around the world. I have not listed them all. This list is not in any particular order. These are the places that left enough of an impression that I still recall having been there. Even though I cannot recall exactly when I visited some of these places, they are still quite vivid in my memory, and clearly recorded and documented in my military records and cruise books.

When I decided to compile this memoir, I decided to start by going back as far as my memory could take me. By doing that I felt I might interest many more people in subjects, which might otherwise be of interest only to me. I'd love to have you come along as I retrace my fascinating journey. I began my trip by taking you back to my childhood. That seems like the best place to start.

Dialing back into my memory, I clearly recall many events from as far back as when I was just two and a half years old. My memory of the events presented in this memoir is as vivid today as if they occurred yesterday. There are, to be sure, more events, which will be recalled as I try to put substantive meat on the bones of this memoir. I am told I was born the son of Collin and Laura Ross Jackson in 1939, who themselves were born just some 42 years after the end of the civil war. I have not located a record of when or where they got married.

My mother was the daughter of Brite and Lillie Belle Ross

and her siblings were brothers Eddie Vance and Fred. Her sisters were Nellie, Loncie and Cecilia. My father was the son of Charlie William Jackson and Daisy Hancock Jackson who both were born shortly after the end of the civil war. My father's thirteen siblings included: brothers Charlie Lee Miles, Claude Eugene, Callaway, William Taft, Fennison Stonewall, John Thomas Henry, Dennis and Booker T. Washington. His five sisters were Dolly Mae, Jewell Odell, Patty Legolia, Gertrude and Annie.

I was fortunate to have known all daddy's brothers and sisters except, Claude, Taft and Jewell. Aunt Patty once made a recording for me detailing a good bit of Grandpa's history indicating he migrated from Mississippi to North Carolina sometime after encountering some sort of legal problems there, changing his name from McQueen to Jackson. He married Daisy Hancock somewhere along the way. She said Grandma claimed to be descended from American Indians, of Choctaw, Cherokee or Creek heritage. I noticed Grandpa had some fingers missing from one hand but he never talked to me about how it happened. The story passed along by family members said it was due to some kind of hunting accident.

****

Of course I don't recall that day I was born but I do recall things quite vividly back to the time I was about two and a half years old. By, the time I reached age nine, my family had moved at least eight times. Seven of those moves are still clear in my mind. We never spent a full year at any one location until 1943, when we moved to Hamp Brewer's farm in the Olive Branch community of Union County, North Carolina. I was almost four when we got there and we stayed until I was seven. Those three years gave me ample time to embark on many exciting adventures. I accumulated plenty of battle scars to illustrate my youthful exploits.

Daddy never talked much about why we moved so much. He just said he wanted us to have the best he could get for us and the sharecropping agreement sometimes were quickly violated due to the houses not being renovated as promised, demands for larger crops than agreed to, disrespect towards us and threats of violence, demands for harder and more work than we small children were capable of performing, and not arranging for provisions at the store as promised. To protest could have resulted in serious harm to us, even death. So it was sometimes better to appear as an Uncle Tom and move on.

Mom and Dad never talked about racism nor did we think of ourselves as poor. We never had a hungry day in our lives thanks to Daddy's ability to hunt, fish, garden, and raise crops and livestock. We may not have liked what was on the table but there was always food there. Sometimes it might only be cornbread, buttermilk and molasses, biscuits and white potatoes or baked sweet potatoes and fatback. Our faded clothes were always clean and mended.

I am told I was born on November 2, 1939, the eighth of twelve children, on Dr. Joe Little's farm in Lilesville, North Carolina. This was a rural farming community near Wadesboro, in Anson County. His farm was known as the "Plantation" and Dr. Little had a dozen or so families working the farm for him. Later, when daddy would take us back there to visit, I used to marvel at the men plowing their mules, twelve or thirteen abreast up and down the long rows, as though performing a well-choreographed routine while the women, wielding their hoes pruned the gardens in perfect unison.

Of course I don't recall that day, but I do recall quite clearly, many events from around the age of two and a half.. In my earliest memories we were living on Sandy Brewer's farm. It was located way back in the woods between the communities of Burnsville and Fountain Hill. We stayed there only a few months, as I recall.

Beginning with Sandy Brewer's farm, and continuing with every farm where we lived afterwards, I readily recall many incidents of significance, which might shed some light on how and why I ended up in the Navy. Some could make for interesting tales of horror, interesting novels or even the basis of an autobiography.

The first house I recall living in was the one at Sandy Brewer. It was big, long and L- shaped with far more room than we needed. The bedrooms seemed a long way from the kitchen where the water bucket was kept on the table. Even at two and a half I was not afraid to trek to the kitchen in the deep of the night in search of a drink of water until that night I encountered the big shinning eyes of the cat sitting on the table near the water pail. I never made that trip to the kitchen alone again.

The rest of the houses we lived in I recall very well. They were all one, two or three bedroom dilapidated buildings except the houses at Dewey Moore and Earl Thomas. Our house at Mr. Dewey was pretty nice with two small bedrooms in the attic, in addition to, the two bedrooms below. Both the house at Lonnie Marsh and Bun Simpson had frightful holes in the walls, floors and ceilings. Daddy kept lime spread around them, he said, in order to keep the snakes from entering.

None of the houses had indoor plumbing or electricity except Mr. Dewey's and Mr. Earl Thomas. Beginning at Sandy Brewer I continued to share beds with my brothers, nephew and other relatives until leaving for the Navy. When the beds were crowded, sleeping at the foot was never my favorite place to sleep.

While we lived at Sandy Brewer's, my father's sister, Aunt Gertrude, gave my cousin Pinky Belle a puppy, and me a pet rabbit. I named the rabbit Peter and ended up stuck with the nickname

"Pete" for the rest of my life. I remember wandering down by the spring one day and finding Pinky's Jack Russell Terrier "Skip" floating dead in the water we used for cooking and washing. We stayed at Sandy Brewer's only a few months after that.

When I was about three we moved from Sandy Brewer's about a half-mile or so up the road to Randolph Thomas's place near Mr. Claude Brewer. We stayed there a short time until his barn burned down. I hated when that barn burned because me and Mr. Randolph's daughter Carolyn, who was about my age, had lots of fun playing there.

I've often wondered if Carolyn had anything to do with that fire. She often pretended she was striking matches. I never saw her with real matches but she loved to put a stick in her mouth and pretend she was lighting a cigar with another stick, or imitating lighting a pipe.

While I was still three, we left Randolph Thomas's place and moved to Charlie Howard's farm near Polkton, where we stayed about three months. While we were there, I enjoyed going into the fields with my father and following him and his team of mules up and down the freshly plowed furrows, delighted by the way the soil felt on my bare feet. We would share the breakfast or lunch that my mother would bring us out in the fields. This was always a real treat because we never knew what the meal might consist of; we just knew it would be good. It might consist of country ham, candied sweet potatoes, fatback, eggs, grits, or even fried chicken with rice and gravy.

About three months, without putting in a crop, Daddy moved on to Lonnie Marsh's farm near Flint Ridge in Marshville, North Carolina. I remember being really fascinated by the way the Border Collies herded his livestock, something I had never seen before.

The way Mr. Randolph and his son, Axel, butchered the animals for market also captivated me. I never understood why the lambs and sheep were so humble as they got their throats slit. They never made a sound other than a simple grunt as they were killed. The young steers that were slaughtered seemed to know what was coming and resisted vigorously. Perhaps it was the smell of the blood in the slaughtering area that disturbed them so much. Whatever it was, it didn't seem to upset the lambs and sheep.

These were really exciting times, to say the least, for an impressionable three-year-old child. Daddy kept us there about five months and then moved a few miles down that fall to Hamp Brewer's Farm at Olive Branch.

This was in 1943 and I was going on five. We stayed at Mr. Brewer's until I turned seven. Until this time we had not spent more than a few months at any one farming location. A lot of things happened while we lived at Mr. Brewer's that affected me in very significant ways.

Mr. Brewer had a grandson who visited with him quite often and was around my age. His name was Charles Coleman. I think he lived in Raleigh at the time. Mr. Brewer doted on Charles and spoiled us both when he was visiting.

The stories Mr. Brewer would tell us about animals and other things as we sat on the porch or hung out in the barn kept us totally riveted and sometimes made us afraid to go to sleep. He told such wonderful "Uncle Remus" stories and would sometimes sneak us into the underground cellar for a few sips of his homemade cider.

Looking back on those days, I realize that Mr. Brewer treated me and Charles like equals. He taught us how to swim and fish in the big pond in the meadow instead of the swimming hole down below the barn. He let us play as much as we wanted in the room he kept filled with toys for Charles.

When we got worn out from playing, he would make us a pallet on the porch and watch over us while we napped in the shade. Sometimes he even let us ring the big dinner bell out in the yard to call the folks in from the fields. We enjoyed that so much that we would sneak off and ring it at the wrong time, which brought the folks hurrying in from the fields from a mile or more away in response to the false alarm.

I never saw Mr. Brewer do any work and he seemed to think our bell ringing was quite funny. The hard-working field hands didn't find it as amusing. Mr. Brewer always took charge of the butchering.

Mr. Brewer's wife, whom we called Mrs. Ada, and his daughter, Ms. Maimie, made sure we always had plenty to eat. Ms. Maimie's hot biscuits and gravy was our favorite. Mrs. Ada's pot roast was a close second and we always had Kool-Aid to drink.

I've often wondered what happened to Charles. When I retired from the Navy, his Uncle Heath said he thought Charles had become a dentist and opened a practice in Raleigh, North Carolina. I have not heard from him since those fun days between 1944 and 1946.

When Charles was not visiting, I would sometimes accompany my mother when she went about a half-mile up the road to help Mrs. Cropsie clean house and grade eggs. That was almost as much fun as hanging out with Charles. Mrs. Cropsie was nowhere as near entertaining as Mr. Brewer, but her son Henry Jay was my age and he had more toys than I could imagine. You name it, he had it. Scooters, sleds, wagons, tricycles, cap pistols and all sorts of other things I cannot name. Best of all, whenever he wanted a new toy he would con his mother into letting him give me a toy or two to take home. Mother didn't get any pay for her help, but she got all of the low-grade eggs she wanted to take home. Our cholesterol was probably off the charts because this went on regularly for more than three years.

During the three years we lived on Hamp Brewer's farm, I learned to read well and was in school before I turned five. No one else I knew around my age could read. I was reading my sisters' fifth- and sixth-grade books better than they could by the time I started school.

During our stay at Mr. Brewer's, I nearly drowned and got bit by a copperhead snake before turning five. Another time, when I was five, my throat was severely slashed by barbed wire while I was chasing Christmas chickens on Ole' Jack the mule.

The snakebite occurred late one summer when I was around five. At that age I liked following my older brother, Eddie, everywhere he went. Hunting, fishing, hiking, swimming, even to work — everywhere he went I followed, if I could.

I started school at Olive Branch when I was four. It was a little wooden two-room building located where the Olive Branch Church now stands. I liked going to school for several reasons. One reason was my oldest sister, Johnsie, and her husband, A. D. Ponds, lived in a house only a few feet from the schoolyard.

Miss Paige taught first- through fourth-graders in one room and Miss McManus had grades five through seven in the other room. Miss Paige's class had four rows of chairs with about eight or ten in each row. She would teach one row at a time while giving the other three rows writing, reading, spelling or arithmetic assignments. From the very beginning, I could read better than the boys and girls in her fourth-grade row and would usually finish my exercises before she had finished giving the other three rows their work assignments.

Often, I was allowed to leave class early and walk the two

13

miles home by myself. I enjoyed doing that in the spring and summer. In the winter, I would stay at school thoroughly bored, unless Eddie got out early to go home and do chores. On those occasions he would let me ride home with him on the handlebars of his bicycle. I remember the feel of the wind rushing briskly across my face.

I remember several of the boys and girls in my first-grade row and became lifelong friends and acquaintances with a number of them. My first-grade row included Warren Reid, who was the tallest, and Ruby Lee Rorie, who could read almost as well as me. Ruby Lee's grandfather, Jimmy Rorie, gave me a dog I named Shag. He was my first dog and became known as the best squirrel dog in the neighborhood. I developed a long-lasting and hopeless crush on Marie Thomas, who was also in our row.

Then there was Coy Hamilton, a boy I chased over three miles one day after he threw a rock and hit me in the head. The Barrino boys, Mutt and Jeff, were in my first-grade row. I can't recall ever knowing their real names because no one ever referred to them as anything but Mutt and Jeff. Incidentally, there were only eight of us in my first-grade class and just sixteen seniors in my high school graduation class, of which I was valedictorian in 1957.

The 16 students in my high school graduation class were:

- Elaine (Bouncy) Davis
- Daisy Ruth Dean
- Yvonne Huntley
- Loretta Knotts
- Vylean Leake
- Ollie Smith
- Harriet Sturdivant
- Elsie White

- Kenneth (Blade) Covington
- Fred Jackson
- Nathaniel (Nae) Knotts
- Louis (Bo Didley) Lee
- Cecil Ledbetter
- Earl Hubert (Hube) Ledbetter
- John Will (Shorty) Little
- Otis (Odaniel) Waddell

Many times while walking home alone from school, the old men sitting around Fred Staton's store would stop me and pay me a nickel or a dime to entertain them by doing the rabbit, or buck dance, sometimes both. Doing the rabbit dance required me to get down on all fours and do a little hopping jig while chanting out "dika dink, dika dink, dika dink dink dink."

The buck dance involved various little tap dance sequences done while singing any ditty I could think up to match my little jig. I usually did the buck dance to the tune of "Oh Susanna." Sometimes I made up my own rhymes. I thought I was pretty good because the men always laughed and stomped and paid me promptly, and the next day I would have money for a Nehi drink and a pack of nabs if I did not spend it right then. That went on for a good while until father stopped me from walking home alone. To get spending change after that, I started raiding Mr. Fred's hen's nests and selling the eggs back to him, two for a nickel.

****

My brother Eddie was my idol, and we enjoyed hanging out together. I remember one particular day he took me swimming and splashing in the creek down below the barn.

After splashing around in the creek for a while I climbed out and proceeded to sun myself while standing on a rock on the bank. I was just standing there, rocking back and forth, when I disturbed a copperhead snake underneath the rock. The snake promptly struck me three times on the top of the right foot and hung on with the third strike. I was able to shake it loose and took off screaming and running toward the barn. I collapsed before I reached it. My brother ran to the rock, killed the snake, then ran up to me and made a tourniquet with my shirt.

Eddie placed me on his back and the snake on a stick and he carried us quickly to the house. When we got home, my oldest sister Johnsie and my mother were in the back yard preparing vegetables for canning. I was fading in and out as Eddie laid me on the porch and went looking for my father, who was working in a field a good ways from the house.

While Eddie was gone, my sister became alarmed at the swelling in my lower leg and removed the tourniquet. This allowed the swelling to progress up my leg. When my father saw the swelling in the leg, he jumped on a mule and went looking for Mr. Brewer, who had a car and could get me to the doctor in Marshville about eight miles away.

When my brother and father returned, they loaded me into Mr. Brewer's grey 1937 Chevrolet, which my father later bought, and we took off for Marshville. I was awake all the way but still in considerable pain.

When we entered the doctor's office, my father was carrying me and Eddie was carrying the snake. Dr. McCleod immediately identified it and sat me down in a chair facing a wall decorated with pictures of different horses. He told me to keep looking at the horses while he worked on my foot. He did something to my foot that hurt like the dickens, and put some sort of medicine and bandages on it. Then he gave my father and Mr. Brewer some instructions and sent us home.

While recuperating from the bite of the copperhead, I was often surrounded by three little people who would appear from behind the furniture. These little folks seemed to be about three feet tall, as was the radio. I suppose they were figments of my imagination, resulting from my feverish state. They sure seemed mighty real at the time. They also seemed very friendly. The little people always appeared dressed alike and I never noticed whether they were boys or girls. They would play with each other as they danced around my bed. Every time I started to drift off to sleep, one of them would grab my right arm and shake me back awake. It was always my right arm. I enjoyed watching them, but I wished they would let me sleep. But they insisted on keeping me awake.

The funny part was they only came out to play when I was left alone in the room. They would immediately disappear behind the furniture the moment someone entered the room. Their favorite hiding place seemed to be behind the big battery-powered radio standing against the wall. Sometimes they would disappear behind the couch.

I don't know where Daddy got that big radio, but he came home with it one evening in time for a Joe Louis boxing match. That was about all he ever listened to. I liked the country western music shows and prided myself in learning all of the words to the songs by Cowboy Copas, Bob Wills, Gene Autry, and the Sons of The Pioneers. "Shoo Fly Pie," "Oklahoma Hills," "Tumbling Weeds" and "Don't Fence Me In" were some of my favorites.

I never found out where that big old radio came from, but we kept it for nine years or so. It must have been some sort of a hand-me-down substitute in some sort of a sharecropping arrangement. Daddy sure did enjoy the boxing matches and, later on, the Grand

Ole Opry and Eddie Arnold on "The Duke of Paduka" show.

The first time Eddie took me to a cowboy movie; I came out of the theater and went around back to see if I could find out where all those horses and Indians went. I thought maybe I'd see the little people back there somewhere.

That old radio finally gave out in 1951. We got a smaller one that sat on a table. I spent hours listening to "The Lone Ranger," "Uncle Remus," "The B-Bar-B Riders," "Jack Benny" and many other short episodes, which were broadcast after school and on Sunday Evenings.

I loved listening to "Little" Jimmy Dickens, "Grandpa" Jones, and the gravelly voice of announcer Grady Cole. My mother's favorite was Clyde McClain, the weatherman on WBT. She would listen to him early in the morning while getting us ready for school. WBT was the main station we could pick up. Sometimes, after we moved to Mr. Bunn Simpson's farm, it was the only station we could get.

One night not long after the snakebite, the little person who seemed to be the leader whispered into my ear that "we have to go now, you will be all right. We won't come back again, but you must never tell anybody." The very next morning my oldest sister, Johnsie, got me out of bed for the first time in days and took me with her to the well to get water.

I know now those little people could not have been real, but they sure seemed like they were. Even now, some seventy years later, I still find myself wishing those little rascals would appear again and tease me, especially when I am having a bad day.

That snakebite was a mixed blessing even though my foot stayed sore for a long time. I was extremely pampered. As a result, I kept complaining and walking on crutches long after the pain had subsided.

That snakebite left me with a lifelong phobia of snakes and memories of the little people I would always see dancing around my bed when I was recovering. Those little people laughed, talked, and encouraged me to stay awake.

This is the first time I ever mentioned them because they told me not to. As a child, I would not dare betray that trust. I remember them as if it was yesterday and I don't believe I will ever get over my fear of snakes.

****

CHAPTER 2

More Fears

"Fear of ideas makes us impotent and ineffective."

-William O. Douglass

Going for Christmas chickens had become somewhat of a ritual whenever mother's youngest brother, Uncle Fred Ross, and his wife would come from Elizabethtown to visit for Christmas. Uncle Fred, for whom I was named, had a real love of fried chicken. Mother tried to provide plenty of it whenever he was around. If he visited anytime during the year other than December, we always had plenty of chickens on hand. For some reason, when they would come during the Christmas season, we always got them from Grandpa over in Burnsville, where he lived on Mr. Troy Lee Edwards' farm. This was usually a fun trip.

****

Mother got a letter from Uncle Fred saying he and Aunt Beatrice were coming in a few days with their two children, Mike and Virginia. Mother decided we would need more chickens than usual since they were staying longer and bringing two extra mouths to feed.

My six-year-old brain kicked into overdrive and I begged Mother to let me go so I could help carry the extra chickens. She thought about it for a brief moment and agreed it might be alright if we took Ole' Jack.

Mr. Brewer had four mules, which the farm hands used to work his farm. There was Mary, the youngest of the bunch, a feisty beast who worked by my oldest brother, Brite Junior. Pet was the best riding mule and the one we preferred to go galloping on. Red Lou was big and evil-natured and very dangerous to be around. He was worked by my brother-in-law, B. J. Rushing, who was married to my second oldest sister, Fannie. Finally, there was Ole' Jack, the oldest and slowest mule of the lot. He was assigned to Eddie for use in the fields.

Even though he was slow and gentle, he was as stubborn as any mule ever was. Eventually, Eddie taught me to plow using Ole' Jack before I turned seven. That turned out to be one hilarious lesson in how not to plow a row straight with a mule while the teacher reclines beneath a shade tree during the entire lesson. The rows I plowed were more crooked than the snake that had bitten me and Ole' Jack turned around in the middle of the field after going only half way to the end of a row. But I got the hang of it. And that was a big mistake. When I was eight, I was given eight hours of work like chopping cotton, cutting cane and pulling weeds.

It was always interesting when the roving blacksmith, the Rev. Clem McDaniel, would come around to shoe the mules and sharpen the plows. He had a way with animals and could get the mules to lift their legs, no matter how skittish they were, so he

22

could fit their hooves properly with new shoes. It didn't hurt that Mr. Brewer's orchard, with all sorts of fruit, was across the road, only a few yards from the blacksmith shop. That made it easy for me to keep the blacksmith, myself and the mules supplied with plenty of nice fruit.

****

The day we were to go for the chickens, in late December of 1945 Eddie wanted to take Pet but knew better than to go against our mother's wishes so we took Ole' Jack and scrambled up on him. I was sitting behind my brother, happy as could be, as we set out on the six-mile trek to Grandpa's house.

When we arrived at our destination, we tied Ole' Jack to a cedar tree, and I set about the business of some serious playing with my cousin Jesse. Late in the afternoon, as we prepared to head home, Eddie hoisted me up on the mule and handed me the reins as he and Grandpa started for the chicken pen to gather up the chickens they had bagged up there.

As they headed toward the chicken pen, Ole' Jack thought he was supposed to follow them and I could not hold him back. He dipped his head low and strolled under my grandmother's clothesline with me frantically yanking on the reins in an effort to hold him back.

Unfortunately, grandmother's clothesline was constructed of barbed wire and latched firmly onto my throat. It barely missed my arteries and I fell to the ground with a hard thud. The mule continued to amble on toward Eddie and Grandpa.

After hearing me screaming and yelling, Grandma dashed out of the house just in time to see me hit the ground. She saw the blood spurting from the wound in my neck and scooped me up.

She stopped the bleeding and bandaged me up really well. Today, I still have very visible scars beneath my chin to remind me of that day.

Not long after Grandma controlled the bleeding, everyone deemed it safe for me to travel and hoisted me right back up on Ole' Jack. Eddie climbed on behind me this time and Grandpa handed him the sacks of chickens. We started for home with me holding the reins, and Eddie holding two sacks of young fryers. Ole' Jack picked up his usual rough, slow gait, as he ambled for home. He needed no guidance as he seemed to know his way.

The first few miles passed by slowly without incident, until we arrived at New Jerusalem Church, about a mile from home.

New Jerusalem was a little wooden white church, which has since been bricked up, sitting in a wooded area with a small cemetery in real close proximity. The sun had just gone down and it was about dusk. I didn't see what kind of creature scurried out of the brush and ran under Ole' Jack's legs. Whatever it was, it startled the calm mule greatly, causing him to panic and spin around 180 degrees in the road and head back toward Grandma's house. I went flying off and somehow landed on my feet, still clutching the reins with all my strength. Eddie went flying another way and both sacks of chickens sailed off in opposite directions, landing on the ground. One sack flew open, freeing the chickens, which darted for the little cemetery.

Imagine a five- or six-year old with a huge gash in his throat struggling to control a panic-stricken mule and lead home in a direction he decidedly did not wish to go. Meanwhile, picture Eddie, stick in hand, chasing the chickens for dear life, knocking them out one by one as they hid among the tombstones in the rapidly darkening cemetery.

24

Eddie successfully corralled the loose chickens and I finally got the mule to follow me back to the church, where Eddie was waiting with the sacks of chickens. Ole' Jack would not let us remount with the sacks. So we walked the rest of the way home. As we trudged uphill the last mile to the house, I was leading Ole' Jack and Eddie had both sacks of chickens slung over his shoulders.

I was very tired, sore all over, and eager to get home. Eddie was worried silly, fearing that the chickens he'd knocked out were all dead. He figured there was a good chance he might get a real good whipping. We made it home, however, without any further incidents.

This story could have and probably should have ended here, but it didn't. The following year all four mules got burned up in a barn fire, caused by a lightning strike or spontaneous heat combustion. It was a fire I will never forget. Even today, some 70 years later, I am still sometimes haunted by the memory of the mournful wailing of those mules as they tried to escape from their blazing stalls. We stood helplessly across the road with Mr. Brewer and his wife Mrs. Ada, tears streaming down our faces, as we watched the burning barn come crashing down on their bellowing livestock. My oldest brother, Brite, frantically tried to chop an escape hole in the side of the barn for the mules to escape. But all four of the mules' stall doors opened into the hallway of the barn, where the ceiling was already crashing in.

Mr. Brewer replaced those four mules with two others named Maude and Kate before we left for Mr. Bunn Simpson's farm. I had a sister named Maude, who hated her name after Mr. Brewer bought that mule named Maude. She never referred to herself as Maude after that.

The Christmas chickens part of this memoir was written December 11, 2012. Eddie died three days later. He never got to read this or any of the other stories he played such a significant role in inspiring me to write about our childhood in rural North Carolina.

In January 2013, I returned to the little country church where the chicken incident took place. Everything appears pretty much the same as it did 69 years ago. The cemetery has expanded considerably and moved from the right side to the left side of the church, and the little wooden church is bricked up now.

The road we were on at the time has been paved to just beyond the church. It is still unpaved from the bridge at the bottom of the hill all the way up the hill to Mr. "Jap" Turner's old house, which is vacant and in considerable decay now. The name J. Turner is still legible on the mailbox post. I often wonder whatever happened to Mr. Turner and his son Ralph, whom my brothers used to help work.

The waves of nostalgia that swept over me as I drove past the church where we were thrown off by Ole' Jack were powerful. It was hard to believe everything had remained practically the same after all this time.

CHAPTER 3

Additional Fears

"The first duty of man is that of subduing fear."

- Thomas Carlyle

The month I turned seven, we left Mr. Brewer's place and moved to Bunn Simpson's farm, also located in the Olive Branch community. Buck Williams, a white farmer living not very far from us, shot and killed my dog Shag soon after we got there. I believe it was because we would not sell him the dog, whose prowess as squirrel dog was well known.

My father replaced my dog by giving me a young calf that was born on a cold February day while I was in school. I discovered him when I got home and immediately named him Mickey after a calf I had seen in an old western movie Eddie had taken me to see. Even though it was a freezing afternoon, I stayed in the shed with Mickey and his mother, Lady, until I was forced to come inside by my mother. Mickey was my constant companion for the next year and a half.

As Mickey and I grew older, I rode him everywhere in the county, including school. We were inseparable until he was butchered by my father and Mr. Vance Braswell, the "fox dog" man who lived up on the hill not far from us. He helped Daddy butcher Mickey in exchange for scraps of meat he could feed the fox dogs he was raising.

Mr. Braswell raised the dogs for sale. Sometimes late at night he would let them out to train, allowing them to run all night long. On occasion he would come down and sit in the yard with my father and listen to the dogs run, identifying the sound of each individual dog's bark.

"Listen Collin. Listen at old Blue go."

"That's Queenie there. She's hot on that old fox's trail."

"There go ole' Rascal, squalling 'cause the pack's done run off and left him. Sure won't get much for him 'lessen somebody's in bad need of a trailing dog."

****

While we lived at Bunn Simpson's place, Daddy was able to accumulate some livestock of his own, plus Mr. Brewer's old 1937 Chevrolet, and a few old farm implements before we moved on. He acquired a couple of old horses we named Bonnie and Tony, and a couple of milk cows we named Sook and Lady. Lady was the Guernsey that gave birth to Mickey.

His best acquisition was a young colt named Nell. It was part quarter horse and the fastest thing around on four legs. I loved watching Eddie beat the other fellas on their old plow horses after they dared to challenge Nell to a race. It was fun riding double on her with Eddie to and from the corn "shuckings" at the Barrett's, Barrino's, Maske's and the Reids under a bright autumn moon.

I always fell asleep clinging to Eddie after all that food, but somehow I never fell off that horse. Nell's gait was so smooth and comfortable.

Later on my oldest brother, Brite, worked Daddy's two old horses to death for a logging company. Daddy was forced to make a plow horse out of Nell along with an old palomino mule named Dan. That was a shame. Nell was never meant to be a plow horse and the palomino mule was horrible to work with in the fields. That's probably why Daddy was able to get him for just a few bales of hay and a few sacks of grain.

While Daddy was accumulating livestock at Bunn Simpson's place, I spent a great deal of time doing some 'cumulating of my own. I was attempting to find a replacement pet for Shag, but they all died or ran away after a few days in my custody. I tried making pets out of tadpoles, frogs, fish, rabbits, squirrels, fireflies, June bugs, turtles and just about any other small critter I could snag except, of course, snakes. My baby sister Gladys would snag snakes and come running to me with them and I would make her snap their heads off, which she would do by grasping their tails and snapping them like a whip, banging their heads on the ground.

Mickey was supposed to be a pet but he was not exactly what I considered a good one. Even though I could ride him whenever and wherever I wanted to, he would also throw me to the ground whenever and wherever he wanted to.

I liked riding Mickey to school and seeing the amazement on everybody's faces. School was a lot of fun for me. I especially enjoyed the morning devotionals when Miss Paige, my teacher, would try to teach us to sing "Row, Row, Row Your Boat," "America the Beautiful" or other fun songs in three-part harmony.

Weekends were full of fun and games like playing Cowboys and Indians, racing the horses up and down the meadow and playing baseball against the white boys at their field on the corner of Highway 218 and what is now Bunn Simpson road. We usually played our games on the rough schoolyard and could only watch the white boys at their nice "field of dreams."

When the news about Jackie Robinson breaking Major League Baseball's color barrier broke, the white boys wanted to challenge our rag-tag team. They even let one or two of us play on their side when they were short players. We were never short players, but we were often very short on bats, ball and gloves.

There were never any fights and the games sometimes lasted from sun up to sun down. The ball field was on land that belonged to Mr. Frankie Thomas and it was a good recruiting ground for him to get help with his farm. He had only two boys and one of them had an arm rendered useless by polio. Still he could play ball and plow a mule with the best of us.

I remember a funny incident one Sunday afternoon when Mr. Thomas's two boys and a few others tried to sneak away with Mickey. I was home with a couple of my sisters. I grabbed the shotgun and went running after the boys, who were leading Mickey away, while my sisters were frantically shouting, "Shoot! Shoot!"

When they saw the shotgun aimed at them they turned Mickey loose and dashed off into the woods. I was all of eight years old and the shotgun did not have a single shell. That Monday me and Daddy went over to talk to Mr. Thomas and he gave his boys a good talking to. They never bothered Mickey again and we kept playing ball on their field until we left for Peachland to sharecrop on Mr. Dewey Moore's farm.

It just occurred to me that I never saw my father avail himself of a doctor or dentist ever. Matter of fact, I recall him being sick enough to stay in bed for more than a day only once. That was

30

while we were at Bunn Simpson's place. Grandma came and stayed with us for a few days and tended to Daddy with potions and herb she gathered from the woods and fields around our house. Now that I think about it, not one of us children ever got seen by a doctor before we were grown, except me. That was when I got bit by the copperhead.

The only argument I ever witnessed Daddy and Mother having happened while we were farming at Bunn Simpson's place. Daddy got off the sawmill truck and walked into the yard, where mother was waiting for him with a butcher knife in her hand. As Daddy tried to cross the yard, Mother kept bumping into him with the butcher knife cocked. All of us children were screaming and hollering and hanging on to their legs. Finally, Mother turned away and Daddy entered the house. I never heard them raise their voices at each other again as long as they lived.

Another thing I recall vividly about my father while I was growing up is I never saw him take a drink other than apple cider. And I never ever heard him utter a word of profanity either, unless "dagnabit" counts. This doesn't mean he never cussed or took a drink. It just means I never witnessed it. I never saw him have sex either, yet he fathered 12 children.

How he must have suffered working so hard, and unable to ever see a doctor or dentist. The only medical attention we children ever received outside the home came as we moved through the school system and got shots and examinations from the school nurses.

****

When we moved to Mr. Moore's farm, I met his son, Dewey Max Moore. I spent many long hours playing and working with

Max until I reached age sixteen. We both were strong, hard workers, but he could outwork me, tossing two bales of hay to my one up into the barn loft.

He could never outrun me when we were playing, though he never quit trying. He taught me how to drive and operate the farm equipment and vehicles used by his father and on surrounding neighbor's farms.

Daddy had his longest stay sharecropping at Mr. Moore's farm. We stayed there about eight years, from 1948 thru 1956. Daddy and Mr. Moore got along well and Daddy continued to acquire more assets, including an almost new Dodge pickup truck, which me and some other young heroes conspired to steal and drive to Mississippi to avenge the horrific death of Emmett Till, who was tortured and killed by white supremacists. None of us had a driver's license or even knew in which direction Mississippi was.

The publicity surrounding the death of Emmitt Till profoundly affected us fourteen and fifteen year-old youngsters to the point of inspiring us to conduct heroic actions to avenge his killing at the hands of white supremacists in Mississippi. So we took daddy's old Dodge pickup truck while he was off visiting with one of his brothers and headed off to Mississippi. We tried to get a tank of gas figuring it would get to Mississippi and back. Not one of us had any weapons, driver's license or even knew in what direction to head to Mississippi. So we drove to Tucker's store and tried to fill the tank up on daddy's account. He would not let us have any gas so we rode around for a while and snuck the truck back home before my folks returned.

Daddy's hard work and determination was starting to pay off as he was able to make more money by staying and sharecropping in places for longer periods of time. Sometimes I was required to wear my sister's hand-me-downs anyway.

Prior to our arrival, Mr. Moore had three different white families work his farm for him. They were the John Taylor family, the Leck Jackson family and the Vernon Wright family. By the time we got there, all three of those families had managed to accumulate small amounts of acreage, which they were attempting to work for themselves.

John Taylor and his family seemed to be doing all right. They had just recently vacated the house when we moved into it, leaving a pretty nice garden patch with scraps of fruits and vegetables that we enjoyed, especially the few stalks of popcorn. Leck Jackson was not doing as well. He was trying to raise cotton after two of his three sons moved away and he could not produce much by himself. His last son, Carl Lee, soon left to open a little country store at the junction of Highway 218 and Tucker Road. The store is still there today.

Poor old Vernon Wright was having a really hard time. He had a very small crop, several small children, and his house burned down around the time we arrived. His children were so ragged and tattered they were reluctant to go to school. They were very friendly, however, and loved to play with us whenever they had the opportunity.

I never learned if those white families accumulated their property from Mr. Moore by way of a sharecropping arrangement with him. I suspected they did because their small acreages adjoined Mr. Moore's property, and they all lived within a mile of his farm.

Max was friendly and nice to work with. We maintained a good relationship up to the time he joined the Army and beyond. He would sometimes take me home with him from the fields when we were working and convince his mother, Mrs. Lillian, to let me eat lunch at the kitchen table with him. Blacks and whites eating together was something virtually unheard of back in the 1940s and 1950s. Sometimes on Saturday mornings they would let us watch "Howdy Doody" and a few other shows, like "Hop A Long Cassidy" and "The Cisco Kid" on the TV with their baby boy, Roy Russell.

Max was there when my mother collapsed with kidney failure. He picked her up, loaded her into his truck and drove her to the hospital in Monroe some fifteen miles away. I was doing duty at the Preventive Medicine Unit Two in Norfolk, Virginia, when I received that news from home. As Max and my mother were on the way to the hospital, someone used our white neighbor's phone to call me in Norfolk.

****

Whenever Max and I were not working, Black Jimmy and sometimes my baby brother Bob, and my nephew, Leon, who we called "Foot," constantly explored the fields, woods, streams, and abandoned buildings in the area. Our forays became even more frequent after Eddie got drafted and sent off to war in Korea, where he stayed an entire year. His absence resulted in black Jimmy and I spending more and more time together.

Believe it or not, I was proud of Eddie when he went into the Army. I thought he looked mighty sharp in his uniforms. His absence did not affect me much because I had already become somewhat weaned from him as he had gone to Charlotte to work at the Purina Plant on North Tryon Street near where the Amtrak station now stands. He was working there and hanging out with two of my older cousins, Theodore Roosevelt and Franklin Delano, sons of my father's brother, Fennison Stonewall Jackson. He was only home occasionally on weekends during the entire year before he got drafted and went to Korea in 1951.

Eddie had given Black Jimmy to me just before we left Mr. Bunn Simpson's place. I originally named him BeBe because he was so small. Before long, he learned how to unlatch the kitchen door and I renamed him Jimmy.

One place we liked to explore was the cattle farming area Mr. Moore maintained through the woods about a mile away from our house. There were two abandoned houses there with fruit trees, scuppernong vines, blue grapes and a pretty nice swimming hole. I have to confess I was quite intrigued with those two old houses, which were about a half-mile apart. There was no evidence of any roads having ever led to them from any main road. There were just old paths and wagon trails which seemed to have led to them.

I was reluctant to explore one of the houses because it was surrounded by thick bushes. Due to my disdain for snakes, I never got any closer to it than the big pear trees in the yard. The other house was located near the swimming pond and had barely any brush or growth on three sides. I suppose the cows had a lot to do with keeping the growth down.

I was able to get up close to it and look in through the windows and one door. Peeking through the door, I was attracted by the sight of treasure and could not resist going inside to explore further.

Inside the room were piles and piles of books and magazines! To me this was truly a wonderful discovery because I had been a voracious reader since I was around four years old. There were no libraries and no money for subscriptions. It didn't matter to me that most of those magazines dated back to the bombing of Pearl Harbor and some, along with the Readers Digests, back even further.

There were Readers Digests, Look, Life, Saturday Evening Post and various other magazines and almanacs. I liked them all but my favorites were Readers digest with its wonderful short stories and Life Magazine loaded with pictures and stories of the fighting from world war two. My baby brother Bob and my nephew "Foot" helped me take home as many items as we could each time we visited the cattle farm. It took us all of one summer to carry all those reading materials home, where I stashed them in the hayloft and took one or two at a time into the house to read.

When the lights were out at night, I would grab my trusty flashlight, which was a bonus for my sales of subscription to Grit papers, and sneak one or two magazines under the covers. I read until I fell asleep or until my parents caught me.

Eddie's homecoming was an unforgettable experience. While there were still two days until he was due home from the Army, I could not sleep and sat beside my parents' bed at two o'clock in the morning in the September cold, shivering in my night clothes and assuring them Eddie would arrive early. They tried but could not convince me to go back to bed. Less than an hour later, Eddie strolled in with his duffel bags in hand, while my parents stared at me in utter dis-belief. I never experienced a premonition as real as that again.

****

I enjoyed the stay at Mr. Moore's. I liked school and we were finally able to ride on a school bus. The first school I attended while we were there was the two-room school in Peachland. It had a huge blackboard separating the two rooms. The blackboard could be raised and lowered so that both rooms could view the stage on our side of the building.

My sister "Stewdie" was always entrusted to carry all of our lunches in one bag and dole everything out to us at recess. Daddy knew how to make the best doggone sausage, souse meat, liver mush and cured country ham. Mother made us treats like apple turnovers and sugar cookies that we called "tea cakes." On holidays she would outdo herself by making cakes, pies and puddings. She always tried to make each of us our very own favorite cake.

Sometimes we had only jelly biscuits for lunch. Most of the time, we had sausage and eggs, or ham and eggs biscuits. Mother always made sure we had at least one or two biscuits each. The toughest part of going to the school in Peachland was trying to concentrate over the loud whining noise of the nearby lumber mill.

Eventually, after reaching high school, I became the bus driver and earned a good driver award. The award was forfeited when I accepted the blame for overturning the bus, even though my relief driver, David Gaddy, was actually responsible for that mishap. I never told anyone about the bus incident except my agriculture teacher, Ben Morgan. Mr. Morgan's only comment was "no greater friend is there than one who would lay down his life for a friend." I don't think he ever told anyone because I have never heard anyone speak of it to this day. Mr. Morgan was full of good quotes, many of which I still remember and have tacked to my walls. One of his favorites and mine is: "A wise man knows how little he knows."

I left Peachland School and transferred to Polkton Elementary School in the sixth grade. The first thing I remember from that school was being assigned to do a book report on Gen. Robert E. Lee's Horse, Traveler. I had never done a book report before and thoroughly enjoyed researching Gen. Lee and his famous horse.

When I was in the seventh grade, the elementary school burned down and my class was required to attend school at the nearby church. It was the same church where I would give my valedictory address in 1957.

During the years we spent farming in Peachland at Dewey Moore's farm, I was able to go to school most of the time.

Sometimes I could only attend intermittently due to farming requirements. I didn't mind farming. I kind of liked it, but I hated to miss out on the leading parts in plays and programs if I could not be there. I hated hearing the jeering of the kids as the bus roared by while I was plowing in the fields.

It also was no fun having to wear my sister Maude's cast-off shoes and coats. It didn't really bother me that they were girl's clothes, but they never fit quite right. I understood what Daddy was going through just trying to keep us fed and clothed. Somehow he maintained his dignity and quietly and steadfastly provided for his family.

I silently vowed that the first thing I was going to do when I got rich was to do something nice for Daddy and Mother. Of course, I never got rich, but I did a lot of little things over the years like providing a new stove, a television or a used car or a few extra dollars.

The other school kids never poked much fun at my clothes. They didn't seem to notice my clothes and usually ended up teasing me more about being the comic book kid. Because they always ended up needing me to help them with their schoolwork, they usually treated me fairly good as long as their girlfriends didn't pay me any attention.

****

I remember when I was in the eleventh grade, my best friend Earl Ledbetter and I got assigned school buses to drive. There was a big tree stump in the middle of the bus parking area and Mr. Sinclair, our principal, was given permission to dynamite it. So he gave Earl his car to drive and asked me to ride with him to Brown Creek Correctional Facility to get the dynamite.

38

He didn't tell us we were picking up dynamite. He just said we were to see the gate guard and pick up a package he had for Mr. Sinclair. Had I known it was dynamite we were picking up, I am certain I would not have accompanied Earl. The dynamite got the job done and our school expanded to several more buses. The extra buses were needed because many of the small outlying elementary schools were beginning to close and send their students to new consolidated schools.

****

After returning from Korea, Eddie was only able to stay home for a couple of days before having to report to Fort Jackson, South Carolina, to muster out of the Army. After his discharge, he gave me his shiny combat boots and stayed around home for a few weeks to teach me the rudiments of Army marching drills before leaving for Charlotte, North Carolina, in search of work. Me, Jimmy, Bob and "Foot" returned to the business of roaming, hunting and exploring all around the county together.

Black Jimmy had come along as we were moving from Bunn Simpson's place to Dewey Moore's farm. Right from the start, he always seemed to think more of me than he did of himself. Jimmy was dedicated, loyal, and fiercely protective of me. There was no creature he would not challenge in my defense. He was gifted with fantastic reflexes, along with natural instincts, possessed a fierce tenacity and was completely fearless.

I watched him take down many a large snake by shaking it to death without getting struck a single time. The most famous of Jimmy's exploits occurred one Saturday evening while we were strolling along the creek bank. We surprised a polecat digging in the leaves. It chose to challenge me instead of Jimmy, which turned out to be a fatal mistake for the polecat.

Before the animal could reach me, Jimmy struck like lightning and grabbed the animal behind its ears and refused to let go until the poor critter was dead. Even then, Jimmy kept attacking every few minutes to satisfy himself the polecat was not going to get up and try to run away. I had never seen Jimmy as furious as he was at the scent of that thing. I tied a string around its legs, tossed it over a tree limb and taunted Jimmy, who reacted fiercely every time I swung the animal past his head.

Jimmy was irritable for days. Every time we started in the direction of where he had killed the polecat, he got overly excited, ran and jumped around like he was crazy. He would roll and twirl in the grass as if he was trying to rid himself of all vestiges of that creature's foul odor.

During my second year in the Navy, Jimmy lost his life chasing a possum into a hollow log. He got his head wedged in the log and could not figure out how to extract it. Bob and "Foot" later found Jimmy's body, wedged in the log, with the possum still scrambling around inside. Had he just turned his head a little bit to the right, he could have slipped out easily.

By this time, Jimmy was more than ten years old, making him the equivalent of a seventy-year-old human. He died as he had lived, fearless and determined. I deeply regret not having been there for him, as he had so often been for me so many times during my youth.

Eddie had brought me that little jet black mongrel puppy when he was about six weeks old, hoping it would help me forget about my calf Mickey, who had been butchered and turned into roasts, steaks and stew meat by Daddy and the "fox dog" man Vance Braswell. At the time Jimmy was given to me, I named him Bebe because he was so tiny. I never imagined what a magnificent and valiant dog that little pup would become.

****

When daddy moved again, it was back to the Marshville area to work Earl Thomas's small farm. I was in my senior year at East Polkton High School and stayed behind with family friends in Peachland so I could graduate with the kids I had been in class with for more than six years.

I stayed in Peachland from November 1956 to May 1957. It was a really tough time for me, but I managed to endure it and graduated as valedictorian of my class. It was a very tough six months, but I am proud that I stuck it out in spite of losing my school bus route and one other frightening experience.

Just before graduation, Tom Winfield, Lewis Lee, Pelham Polk and I were riding home from a tournament basketball game in J.C. Bennett's car when we were shot at by a group of white fellas we knew as we drove by a service station in Peachland.

As we passed by them, they shouted at us and called us names then fired a gun at us, shattering the rear window of our car. The bullet lodged in the back of the seat where I had been seated about a half an inch from where my neck had been resting.

We raced to State Trooper Gene Dutton's house nearby to report the incident. He came to the door in his nightclothes and we told him what happened. He threatened us with arrest if we didn't get off his porch. He also knew the young white men who had shot at us.

I developed a deep distaste for Gene Dutton, which persists to this day. The lack of respect developed for law enforcement officers where I grew up was subsequently reinforced by several episodes in which law enforcement figures across the nation treated African-Americans unfairly and unjustly. Before I joined the Navy my reaction to unfair law enforcement was that it was

41

expected to be tolerated so as not to bring additional harm or danger to the family. Later I became a little more militant and lost most if not all, respect for uniformed policemen. Finally, I did lose all respect for cops and have not totally regained it as I have continued to witness the ongoing unjust actions like that imposed on Rodney King, myself and many others.

Later, I was told that Gene Dutton's best friend, Tommy Sykes, had been killed on his motorcycle. I was stunned that I got a good feeling at hearing the news. Another encounter with local law enforcement occurred when I slid into a ditch on an icy road. This happened about a hundred yards from where I lived on Mr. Randolph Thomas's farm when I was three years old. I walked up to their house and called the Highway Patrol. I wondered if the woman who let me use the phone was Carolyn, with whom I used to play so many years before.

Anyway, the state trooper responded, arrested me on the spot and locked me up. When I got to the court in Wadesboro, there were sixty-five cases on the docket. Sixty-four of the defendants were black and one was white. Gene Dutton and other associates of Tommy Sykes ticketed most of them. I was found not guilty but they could not find my driver's license. I asked the judge about my license and the disparity in ticketing and he threatened me with contempt of court.

Another time, while I was home on leave in 1960, I was arrested and locked up for waving at two white girls who had waved at me first. I was charged with disturbing the peace, held overnight at the jail, but let go the next day when my father and Eddie showed with bail money. I was released and told to return to my base in Philadelphia and warned never to set foot in North Carolina again.

I am sure my life was spared that night by what transpired during the arrest. My car had been surrounded by several

42

policemen, sheriffs, and state troopers and deputized vigilantes as I was leaving my grandfather's house in the Viney Bottom around midnight. I had New Jersey tags on my car. I had Bobby Thomas and Queen Clyburn riding with me and the policemen asked me to get out of my car and come with them. As I was getting out of my car, memories of the Emmitt Till incident rushed over me and I felt this was going to be my last day alive. So as I got out of my car I said to Bobby and Queen in a voice loud enough to be heard by all the people surrounding my car," take my car to my daddy, Collin Jackson over at Mr. Dewey Moore's and tell them to call my commanding officer at once.

There is no doubt this gave them pause, the fact that they had a military person in custody with two witnesses. My father was well known and respected throughout the county and so was Mr. Dewey. There is no doubt they wanted me to try escaping because they directed me to ride in a car with just one deputy who took his gun out of its holster and placed it between us on the front seat. As he was driving us away from my car I asked him what were they arresting me for and what were they going to do with me? He simply said, "shut your mouth nigger and you will find out. When we reached Polkton where the "waving" incidents had taken place, the deputy left me in the car alone with his gun and went upstairs to get some papers from the Justice of the Peace while several of the plain clothed vigilantes waited near his car. He soon returned and we proceeded on to the Jail in Wadesboro.

I was sorry to feel a sense of satisfaction when I heard policeman Tommy Sykes had been killed on his motorcycle. However, reflecting on the unequal and unjust treatment I received from law enforcement and the court system where I grew up was normal. My disdain and lack of respect for racist officers was strongly reinforced by what was happening throughout the South with dogs and fire hoses at that time.

The next day we reported the shooting incident in Peachland to the sheriff in Wadesboro and they charged the men with assault. A pre-trial hearing was set. At the hearing, court officials convinced everybody except me to drop the charges and the case never went to trial. The magistrate or prosecutor threatened to charge us Nigras with inciting a disturbance if we did not drop the charges and the case was dismissed without ever being heard in court.

A few years later while I was home on a weekend pass I heard from my cousin Jesse Ross. He stood on a corner in Wadesboro and witnessed the death of Officer Thomas Sykes when a drunken driver demolished the motorcycle he was riding. At the time I was an advanced medical corpsman authorized to perform duty on board ships without a medical officer. I wondered if I would have tried to help the officer if I had been on that corner. Because he was such a good friend of Trooper Dutton's and had a reputation for being brutal toward Negro folk, I'm not sure what I would have done. Maybe my training would have kicked in. I'm glad I've never been faced with a dilemma like that.

There are so many instances I recall of being mistreated by law enforcement officials in many states during these years. I won't dwell on them hear because I get depressed when I think about it.

Mr. Moore's large farm was becoming too much for Daddy to handle with his shrinking labor force so he moved to Mr. Earl Thomas's place near Marshville. He was farming there when I left on my great adventure in the U.S. Navy. That journey spanned 21 years and included visits to five continents.

****

After graduating from high school, I went to Jersey City, New Jersey, to spend the summer and worked in Union City. There I worked at GolPak Corporation's meatpacking plant. My goal was to earn some money to pay for college.

My parents were proud of me and sacrificed tremendously so I could finish my last school year at Polkton. They were very elated about my coming in number one at the head of my class of sixteen, but they had no idea what an academic scholarship to Shaw University was and had absolutely no money to help me pay for college. So college was not an option because I had also lost all my good clothes in Roanoke, Virginia enroute home from Jersey to start school.

On the way home from New Jersey to attend college, all of my clothes fell out of my brother Brite's car in Roanoke, Virginia, while we were crossing a railroad track. An address of a friend was discovered in my lost clothes and the clothes all returned to me several months later.

When I got home from Jersey, I asked Mr. Thomas if he would give my father a cow to sell, so he could help me pay for college. I had won an academic scholarship to Shaw University in Raleigh, North Carolina, but had no money or adequate clothes. Mr. Thomas said he could not afford to give up a cow at that time. Daddy was already planning to move back to Mr. Dewey Moore's farm. I don't think Mr. Thomas was happy about the fact I had not come to his farm to work after graduation.

Daddy was by no means a coward, but I did not realize how much courage was required to do what he did for the sake of his family's survival, until my trip was practically over.

****

While I was preparing to start this journey, Daddy was waiting to return to Peachland and Mr. Moore's farm for a second time for

45

his final attempt at sharecropping. He'd had some success there before and Mr. Moore was not looking for such a large farming operation that would require a lot of manual labor because his son Max had moved on and so had I. Daddy was accompanied by my mother, who later passed away there, the two youngest of his twelve children, Bob and Gladys, and one grandchild, Barbara Ann.

Now let me get on with retracing the challenging road my military career took.

****

In December of 1957, I spent a several weeks riding around with a couple of friends, Gene Thomas and Herbert Louis Tillman, looking for work. We had run out of money and people to lend us their cars. As a last-ditch effort, we decided to enlist in the U.S. Navy on the buddy plan. This plan guaranteed friends who signed up at the same time the opportunity to go through boot camp together and possibly get assigned to the same type duty for the entire enlistment.

We headed down to Wadesboro to take the enlistment exam administered by Recruiter Caldwell Ross. He was eager to meet a quota at the end of the year and allowed us to take the exam together while he waited outside with a cup of coffee. I tried my best to give Gene and Herbert the right answers. Caldwell scored our tests and said I had one of the best scores he had ever seen, but Gene and Herbert failed miserably. Caldwell said to me, "Congratulations! You are in, but unfortunately, your buddies won't be accompanying you."

Gene and Herbert confessed to me years later that they had deliberately failed the tests and had never intended to enlist. They

46

had hoped I would flunk too. Over a period of several years, we discussed their reasons for not wanting to join the Navy. I think they eventually regretted that decision.

Gene became a motorcycle mechanic and Herbert became a Baptist preacher. They had promised to come see me off whenever I was to leave for the Navy.

On the chilly morning in January 1958, I waited with my father with no sign of Gene or Herbert. A few sawmill workers waited with Daddy, hovering around a warm fire burning in an empty metal drum and me. I kept looking for Gene and Herbert, but they never showed. They told me later that they could not get a ride.

The sawmill workers were waiting for their work truck, and I was waiting for the Trailways bus. We were waiting beside Highway 74 in Marshville, North Carolina, on the very spot where a Stegall Smoked Turkey Store stands today. This is directly across from the store where, many years later, Oprah Winfrey was knocked to the ground in the movie "A Color Purple."

PART TWO

The Search For Adventure And Enlightenment

****

USS GUAM (LPH-9)

****

CHAPTER 4

Ongoing Fears

*"The free man is he who is not afraid to go to the end
of his thought."*

- Leon Blum

The bus finally arrived around sunrise and dropped me off in Hamlet, North Carolina. In Hamlet, I boarded a train for the trip to the Armed Forces Induction Center in Raleigh, North Carolina. On the train I started to feel sad as I realized that I was actually leaving home. The rocking and swaying of the train soon lulled this 18-year-old farm boy to sleep. I gradually dozed off, thinking of the folks being left back home and wondering what in the world the future held in store for me.

I spent two days getting processed at the induction center in Raleigh. I was billeted in a seedy YMCA a few blocks from the center. My white companions were lodged in a nice hotel across the street from the center. It had been ten years since President Harry Truman signed the presidential order banning segregation in the armed services, but Jim Crow was still very much alive and well in North Carolina and would travel deep into the military with me.

I had no way of anticipating how serious institutional racism, bias and bigotry were in the military, or how dramatic an affect they would have on my efforts to excel. I still felt like segregation and discrimination were ordained by God, legal statutes, and rightfully imposed on us Negroes. Such was the nature of my cultural conditioning and racial awareness. I was sworn into the Navy on January 28, 1958.

****

Upon completion of the induction process, our records were given to the smallest man in our group, a young fellow named Agnew, to transport to the Naval Training Center in Great Lakes, Illinois. We boarded a twin-engine airplane at Raleigh-Durham airport and I settled in for what was to be my first of many very scary airplane rides.

Bucking fierce headwinds, we flew to Washington, D.C., where we changed to another plane that flew us nonstop to Chicago. We landed safely after a really rough ride in which the plane dropped as much as a hundred feet at a time as it hit deep air pockets.

On the ground, a driver in an old, chilly bus met us during the worst snowstorm I have ever witnessed. He drove us to the receiving center at Great Lakes Naval Training Center, where we watched Elvis Pressley in the movie "Love Me Tender" before getting assigned a place to sleep.

Being the only black man in the group, I rode to the base in the back of the bus, where there was absolutely no heat. Coming from the segregated South, I didn't think I could ride up front where the meager heat was being generated. The right to ride in the front of the bus was still not fully an acceptable way of life even though Rosa Parks had made her stand in 1955.

****

On January 29, 1958, at 0500 hours, I was awakened at Naval Training Center in Great Lakes, Illinois. Never in my life had I seen so much snow! Little did I know it would snow almost every day I would be there for boot camp. The bitter cold was quite a shock to a skimpily clad southern youth used to far milder temperature.

The weather was not nearly as shocking as my first military meal, where I was introduced to three eating utensils to be used at the same time. A spoon, a fork and a knife too! What the hell was the knife for? I was a high school graduate, and a valedictorian at that, so I just looked around to see what the other folks were doing with their knives and got the hang of it right away. Welcome to the world of cultural shock and indoctrination into the military world of white folk!

****

Standing watch on an empty clothesline has yet to make any sense to me — particularly in sub-zero weather.

During classification I wanted to become an engine mechanic, storekeeper, or a personnel man. I was a pretty good typist at about sixty-five words per minute and hoped I would at least have a shot at being selected for some sort of clerical job. But waiting outside in the cold and snow for forty-five minutes before taking the typing test seriously hurt my score. As a result, the enlisted classification section assigned me to be trained as a medical technician. Medical technicians were affectionately known throughout the Navy as "pekker chekkers." I was to learn very soon that there were numerous other slang names for a hospital corpsman like "chancre clanker" and "pill pusher." Pekker chekker was by far the most commonly used.

Pekker checking was the dubious exam that medical corpsmen performed on enlisted men periodically. Called "short arm" inspections, they were surprise examinations of the genital area to detect and control venereal diseases. I wanted to be selected for anything but that assignment and broke down in tears when I found out. I didn't get any of the positions I had listed on my preference sheet at classification and, in the end, got the one and only thing I didn't want. I learned later, however, that it was the best possible option and led to many really good assignments throughout my entire Navy career.

Though I loathed the idea of becoming a pekker chekker, I later realized what a far better deal it was to be a medical corpsman than it was to work as a dirty, greasy engine mechanic, stuck in a hot and noisy engine room, or some office flunky with a haughty personnel supervisor constantly peering over my shoulder.

****

Boot camp was a tough period of training conducted almost entirely during snowy weather. I was chosen as the company flag bearer. It was a pretty good assignment that allowed me to always be the first in the company to enter the dining hall and the first into the barracks out of the snow and cold. I landed the assignment by demonstrating the rhythmic cadence and marching skills my brother Eddie had taught me when he returned from Korea.

About halfway through boot camp we were treated to entertainment provided by the Royal Tahitian Dance Troup. Man,

I had never seen anything like that! The men danced and twirled fire, while the women swayed and twirled their grass-skirted hips and the drummers danced atop their large drums. I made up my mind then and there that I was going to get myself an assignment somewhere in the South Seas.

****

After completing boot camp, most of the white guys that had enlisted with me in Raleigh boarded the train with me and we headed home to North Carolina together. We enjoyed a kind of camaraderie that was surprisingly warm and friendly. It was completely different from the palpable disdain I had experienced growing up, at the induction center and in the airports on the way to boot camp at Great Lakes, Illinois. During all of the examination procedures, I was forced to be the last one to be processed. There were dozens of airmen, soldiers and Marines being processed ahead of me at the same time. At the airports, the little guy, Agnew, who was in charge of our records, would not allow me to board the plane until everyone else had boarded and found a seat.

On the way to Raleigh, it was apparent to me that something had changed, although I was not quite sure what. It became even more apparent when we got off the train in Memphis to eat and every white sailor traveling with me refused to eat at the restaurant where I was refused entry. Boot camp had impacted all of our attitudes. I dared to go where I never would have attempted to go before. The white sailors with me exhibited a kind of empathy, which they would never have displayed back home in North Carolina.

****

After spending fifteen wonderful days at home on leave, I returned to Great Lakes for hospital corps training. I never encountered any of the men who went through boot camp training with me since that time. I've often wondered if they continued to modify their attitudes regarding black and white service members serving together. I certainly changed my outlook over the years, both in and out of the Navy.

Hospital Corps school was conducted for sixteen weeks at Great Lakes, Illinois, during the spring and summer of 1958. It was exciting and I thoroughly enjoyed the pleasant environment interacting with people from all over the United States, Germany, Canada and the Caribbean. There were mostly black and white sailors in my class, but there were a few other races as well. We had some Japanese, Pakistanis and a Korean in our class. I took right to the medical training curriculum. The weather was nice during the entire period of training and we enjoyed many cookouts along the shores of Lake Michigan.

There was a brief period when rumors put a real scare into us as combat flared up in Lebanon and our instructors informed us to be prepared to have our formal training cut short and get sent to the fleet. The instructors knew that would not happen, but they also knew it would inspire us to concentrate harder on our training.

Our training was oriented toward facilities such as hospitals, dispensaries, and other clinical health-care facilities. The curriculum contained virtually nothing about combat emergency situations. Our training concentrated mostly on medical situations in support of facilities staffed by doctors, nurses, technicians and specialists. My first after class assignment was to attend an autopsy

58

to introduce us to some of what a medic would be expected to learn about human anatomy. That was a real shocker.

Training in management of traumatic battle injuries would come much later at Field Medical and Advanced Hospital Corps Schools. I knew absolutely nothing, and we were taught very little about traumatic medical emergency treatment procedures at this time.

During Hospital Corps training, I got my first look at the big city of Chicago, where we were allowed to go when we were off duty on weekends. This visit to "Chi-town" thoroughly convinced me that I was not cut out for big city living. I did not know what a gay person was and had never been approached by one before. On my first and only visit to Chicago, I was approached no less than three times by what was known back home simply as punks. The first time was while I was in a movie theater and a man moved into the seat next to me and kept touching me in the side until I got up and left. The next time was at the pinball arcade. While I was playing, a man came up and took my arm and tried to show me how to work the machine. The next time was on the city bus when a man moved all the way to the back and sat down beside me even though there were a number of empty seats all around. It finally dawned on me what was happening, and I've been back to Chicago only once since that time in 1958. That was in 1998 as a civilian, some forty years later.

****

My first duty station after medical training was in another big and challenging city, Philadelphia. I was able to handle Philadelphia better than Chicago because I was learning more about a lot of things that I had never thought about before — things like other people's tastes in music and entertainment and how to conduct myself in different cultural settings.

I had several relatives in Jersey City, New Jersey, about ninety miles away, which I could visit. Several acquaintances from my Hospital Corps School class ended up being transferred to Philly as well. There were a number of relatives and acquaintances from back home in North Carolina living in Camden, New Jersey, just across the Walt Whitman Bridge, with whom I could also visit during off-duty time.

****

In September of 1958, Naval Hospital Philadelphia became my first regular shore-duty assignment after completing boot camp and basic Hospital Corps School at Great Lakes, Illinois. "Good afternoon, sir. Hospital man Apprentice Fred T. Jackson, 523 20 08, reporting for a normal tour of shore duty."

After check-in and orientation, I was assigned to orthopedic ward 3-D. Across from my ward; a corpsman named William H. "Bill" Cosby Jr. was assigned. During late September 1958 and early 1959, I enjoyed some moments on the basketball court and in the chow hall with Bill Cosby. He was rather distant and that was about the extent of our interaction before he was discharged. He never really fraternized with the rest of us very much because he lived in Philly and attended Temple University on his off-duty time. He only hung around the barracks on the days and nights he had duty. He was discharged from the Navy a few months after I reported to Philadelphia for duty. He never displayed the slightest hint of the humor, which would make him famous. Sometime later, while I was fighting for a respectable place in the military scheme of things, I looked up and there was old Bill co-staring in a TV show called "I spy."

As we all know, Bill Cosby went on to bigger and better things in the entertainment arena. He never showed very much interest in the Navy and never rose very high in the ranks, nor did his father, who had also been in the Navy. I doubt if Bill would recall me now, but I've often wondered if he might. Believe it or not, I did get to spend some time later with his father.

Later, while stationed at Naval Hospital Newport Rhode Island, and while I was supervising the X-ray department, Bill Cosby's father was hospitalized there for a while with a leg injury. The older Cosby had been a deserter from the Navy earlier, but his picture, along with his son Bill's, hung in the administrative office at the Naval Hospital New Port with a picture of the Commanding Officer in 1971. It was still there when I left there for duty at Camp Lejeune later that year.

****

My tour of duty at Naval Hospital Philadelphia was, for the most part, quite enjoyable. This was a wonderful period in my cultural adjustment, militarily and socially. Being fresh off the farms of North Carolina, I found my new duty station to be a real eye-opener and a wonderful learning experience. There were quite a few black corpsmen stationed there, but to my amazement, I was the only one that I knew of from below the Mason-Dixon Line. The others I recall meeting were mainly from the West, Northwest or Northeast. Most of the other black guys' views about life seemed to be considerably different than mine. They had come to believe much earlier than I had that discrimination was wrong, and that pushing back was the right way to go. I enjoyed hanging out with them at skating rinks, amusement parks, Atlantic City and nightclubs.

I didn't particularly care for their tastes in music, but I enjoyed the entertainers we would go see at the Showboat

Club or Peps Lounge. Entertainers like Dave Brubeck, Nancy Wilson, Cannonball Adderly and other jazz artists were frequent performers at these clubs. I was more of a folk, country western, and rock and roll enthusiast, but enjoyed watching some of the artist like Dizzy Gillespie and Della Reese perform.

Philadelphia Naval Hospital was located at Broad and Patterson Streets, just across from the sports complex at Soldier Field. It was also just a short walk from the Navy shipyard, Walt Whitman Bridge, and a nice park outside the main gate. The Walt Whitman Bridge allowed quick access to Camden, New Jersey, where my Uncle Dennis and Aunt Geneva lived with their five children, Leonard, Maimie, Dennis Jr., Johnnie Frank and Jewel.

My father's brother, Uncle Dennis, whom we called "Demps," had come to New Jersey with his family from T. A. Jones' farm in Burnsville, North Carolina, around 1955 or1956. They first went to Swedesboro, New Jersey, to pick vegetables, but moved to Camden Proper seeking work at the Campbell Soup plant. It was easy to visit with my cousins in Camden whenever I was broke or didn't have enough time to go all the way to Jersey City to see my five sisters and two brothers who were living there.

My sisters and brothers, Laura Luvenia (Bean), Maude Marie, Dorothy Elizabeth (Stewdie), Flora Fenton (Nannie), Ella (Dosha), whom I would later learn was not named Ella at all, two brothers, Brite Junior (Marsh) and Collins William (Eddie) were all living in Jersey City at the time. I never knew exactly where my siblings got those nicknames. I got my nickname when I was about three years old. My Aunt Gertrude gave me a rabbit that I named Peter. The nickname "Pete" stuck with me from then on.

****

A lot of folks from back home that I knew very well were living in Camden. They included Gene and "Buck" Biles from Olive Branch, Nanny Kate Benton from Olive Branch, Charlie Davis from Burnsville, "Teet" Manus and his three sisters, Lola, Lottie (Tot) and Ruth, also from around Olive Branch. Nanny Kate's two sisters, Cassie and Nooncie, were there, too. I had attended Olive Branch school with all of them except Charlie Davis.

I remember seeing Buck Biles get all of his front teeth knocked out by a beer bottle at the Blue Goose Club in Camden while trying to defend my cousin, Dennis Junior, in a fight Dennis had initiated. Needless to say, Camden was a pretty rough place back in the late 1950s.

Philadelphia provided so many exciting and interesting things for a country bumpkin like me to enjoy, and it was loaded with history, one of my great interests. Admission to the sports complex at Soldier Field across from the North Gate was only fifty cents with a military ID. It was free if you were accompanying a patient from the hospital.

****

The icing on the cake came when my brother Eddie co-signed for me to buy a nice green 1953 Pontiac. This made me very popular because I was one of the few sailors, black or white, who had saved up his money for a car, rather than spend it on fancy clothes. This brought me far more so-called friends wanting rides than I cared to associate with, but we all seemed to get along, except for one sailor, who constantly wanted to borrow my car. I only recall getting into one fight the whole time I was stationed in Philly, and it was with this sailor. Because I would not loan him my car for a hot date, he decided to taunt me in front of a large group of sailors. We were standing around in the lobby of the barracks at the time, waiting for our once a week cleanup assignments. This particular sailor began entertaining himself and the crowd by ridiculing my country background, degrading me as a farmer and backwoods muleskinner. Both the white and black sailors lounging around urged him on. As I turned to walk away, he gave me a hard whack to the right side of my head from behind.

He outweighed me by a considerable amount, but before he could enjoy his brief moment of glory I was on him like my dog Black Jimmy on that old polecat. After lifting him into the air, I body slammed him to the floor and landed on top with my left forearm across his throat and my right hand filled with his scrotum squeezing real hard. He started screaming really loud, but I wouldn't let go. I finally released him after emphatically admonishing him that if I noticed the slightest scratch on my car I would rip his gonads out for real.

I hoped he did not have anyone in the crowd who had a grudge against him and might damage my car, hoping I would retaliate by going after that sailor. But no one ever bothered my car, and the sailor had learned that farmers like me who worked with mules could be strong like mules.

His buddy Reno didn't care much for me. But Reno didn't like his friend enough to risk attacking me either. I never had any more trouble with that sailor or any other fights. He became the butt of many jokes around the barracks for quite some time. The white and black sailors alike teased him often about getting his butt kicked real good by a "country mule." Some of the black sailors I remember who probably witnessed the fight were Robert Reno, Rudy Jones, Tony Anthony, Edison Davis and "Big" Jim Myers. They worked the same shift as I worked and should have been there. After the fight, the sailor tried to become friendly with me,

64

but I kept my distance from him. He was transferred to Naples, Italy, a few months later. My duty at the Naval Hospital continued as a good tour, marred only by a few unpleasant encounters with the Uniform Code of Military Justice.

****

There were several disciplinary incidents in Philadelphia, which were apparently related although I didn't make the connection at the time. About 55 years later I begin to tie it all together after I retired, received my military records, and decided to write this memoir. Only then could I see how all the disciplinary actions against me at Philadelphia might have been related.

The first incident involved the morning nurse on Ward 10 charging me with insubordination and disrespect to an officer. As I came off night duty, she asked me to give the morning report. As I reached for the files, she grabbed my wrist and said, "Nigger, get your black hands off that file and give me the report verbatim." I snatched my hand away and gave the report the best I could without using the files. We were always required to use the files. It was standard procedure and that's why I had reached for it.

Cardex files were used on all the wards and contained a running narrative on each patient for the entire shift. After giving the morning report, I left the ward and proceeded to the barracks to prepare for breakfast. The nurse contacted the Officer of the Day and placed me on report for disobeying her orders and insubordination for removing my hand from her grasp.

Commander Timberlake from the Pharmacy investigated the charges. He held a pre-trial investigation and ruled the charges invalid after two patients who witnessed the incident came forward to present written statements and to testify before him.

A soldier and a sailor whom Commander Timberlake interviewed had witnessed the incident. They were willing to testify on my behalf during his pre-trial review so he dismissed the charges.

I probably would have lost that round and been drummed out of the Navy if those two witnesses had not shown up. I am certain my career would have ended then and there had I not prevailed. I remained on night duty for a while longer but the young nurse from Tennessee was immediately transferred. I think she must have been sent to another base because I never saw her again.

The second incident took place while I was still on night duty on Ward 10. I had occasion to go to our supply room on another floor for some materials. As I entered the dark storage room I happened upon a nurse who was the supervisor for three medical wards, including Ward 10. She was having sex with an HM3 as I tried to sneak out of the room without being noticed but they spotted me.

The Petty Officer (HM3) begged me to keep my silence and the nurse followed me into the pantry on Ward 10's floor. She was crying uncontrollably and pleading with me to keep quiet. She said if I did, she would see that I got the leave I had requested from the Tennessee nurse and a transfer off the contagion ward as well. I did not like working the contagious ward. I had worked there and on other wards long enough to qualify for a rotation into other aspects of the hospital such as supply, transportation or the Master at Arms Force. I had previously been hospitalized for a week with pneumonia contracted while assigned there.

I had been trying to get off that ward for months. The ward nurse supervisor knew I had been trying to get approved for leave and a transfer, which the young nurse from Tennessee had not approved. This commander had always been one of my favorite nurses, and now I felt extremely bad for her. She had at least 12 years of service and her career was now in jeopardy, so I agreed to her terms. She was true to her word and got my leave request approved immediately. That's when I went home for ten days and learned my dog Black Jimmy had been found dead.

My baby brother Bob and my nephew "Foot" were walking through the woods when they found him. On our way to the fishing pond, I asked where Black Jimmy was. Bob said they hadn't seen him in a couple of weeks and that he might have run off somewhere, which was highly unlikely. They said they had spotted an animal skeleton with its head stuck in a hollow log surrounded by scraps of black fur. We examined it and agreed it was Jimmy's remains. They pried his skull out of the log and found a possum still alive inside.

Had I not been such a decent person, there is no telling what goodies I could have squeezed out of the Commander. When I returned from leave, the supervising nurse, again true to her word, arranged a transfer for me to the psychiatric wing of the hospital to work under the supervision of Nurse Hall.

****

Nurse Hall also was a commander. She was extremely nice and a pleasure to work with. I rotated among three different wards in the psychiatric unit for Commander Hall. One of her wards was geriatrics, where William "Bull" Halsey, the famous World War II admiral was hospitalized. Admiral Halsey was a fascinating character. He liked to talk nautical stuff all the time and he tied his pajamas into ridiculous nautical knots. There was no way you could untie those knots. We had to cut his pajamas off to bathe or change him.

I also worked the ward where shock therapy was administered. That ward was only used once or twice a week when shock therapy was administered. It was hard to watch. It seemed to be such a harsh procedure. The patient would shake violently, drooling at the mouth, and eventually pass into unconsciousness. Our job was to restrain the patients and keep them from choking or hurting themselves while the doctors applied the electricity.

I also made the weekly run to McGuire Air Base and Fort Dix, New Jersey, to pick up and escort restrained psychiatric patients arriving from all over the world. Philadelphia was a major psychiatric receiving center for the Navy and we often picked up some very disturbed individuals.

I remember some of the off-duty corpsmen caught a stray cat and administered shock therapy to it one night. For several hours, the cat could only walk backwards. I disapproved and let the other guys know in no uncertain terms the trouble we could all get into if Commander Hall found out. Coming from a farm, I had a great respect for animals. They said I had better keep quiet about it and I did because there were three of them and only one of me. They were also soon to be completing their four years of service. It would be many, many years before I would even think about developing an attitude driven by the kind of vengeance or revenge displayed by so many of my fellow sailors.

I mostly worked the main ward for Commander Hall. It was not a locked ward. Even so, it was kind of creepy to work the night shift alone. The patients would come up to the locked nursing cage and present all kinds of crazy requests when they should have been in bed. One patient continually presented request slips to have intercourse with one of the day nurses. One night Commander Hall was there on her rounds when he brought his request. She calmly took the request and approved it in front of the patient and told him she would keep the request and see if she could get the doctor to approve it too. The sailor went to bed smiling and never brought up the request again. There was another patient who insisted he was a refrigerator and would come to the nursing station asking to be defrosted. Commander Hall was one cool customer. I guess that was why she was supervisor of the entire psychiatric wing.

Another thing I liked about Commander Hall was that she would let us play record albums in the soundproof nursing station after the lights were out on the ward. The records were mainly hers and included the Kingston Trio's complete collection, some Harry Belafonte and a few other folk groups. Sometimes, after I went off-duty, I heard "Hang Down Your Head, Tom Dooley," "The Sloop John B," or "Day-O" playing in my head.

****

During the time I was assigned to the psychiatric unit, someone added my name to a watch list to relieve for chow, after I had already been relieved of duty for the day. When working the morning shift, you could be selected to return at night once a week to allow the night crew to go eat. Your name would be placed on a list for chow reliefs and you had to check the list every time you went off duty to see if you had been placed on the roster for that night. You were never selected more than once a week and you were not selected every week. I was charged, convicted and punished for failure to relieve for chow.

Another time while I was still working for Commander Hall, someone added me to a training roster for a class I'd already completed or exceeded the requirements for. I had already been promoted beyond the requirement to attend. The class was for ranks E-1 through E-3 and I had already been promoted to E-4. Once again I was charged, convicted and punished for failure to attend training.

There were no acceptable explanations because the personnel office only went by whether your name was on the training roster or not. Whether you were required or not required due to any particular circumstance did not matter. The only thing that mattered was whether or not an instructor checked your name off the list. When you were placed on a list, the instructor did not question if you should have been there. He simply did not check you off if you did not show. Only the personnel office knew if you were supposed to attend a particular class.

I've often wondered who put those false watch lists up after the real watch lists had been published. I wondered if someone who liked the young Tennessee nurse, or perhaps the rogue corpsmen were responsible. Whoever did it succeeded in getting me a record of disciplinary actions that would stay with me for the duration of my career, affecting me in ways I would not understand for a long, long time.

It happened twice and I was punished both times. Punishment was quite light, a mere five days restriction to base in each case. I suspect the dismissed charges brought by the Tennessee nurse probably had something to do with that, as well as the fact that if they had wanted to, they could easily have verified the fact that I was innocent.

I really wondered if it might have been done by the rogue corpsmen who administered shock treatment to the stray cat. They were all corpsmen finishing up their four years in the Navy and due to be discharged in a few months. They didn't care much about rules and regulations as long as they could get away with their wild stunts. They were what Navy men referred to as "old salts," meaning they had been in the Navy for a long time and knew their way around the military system.

In spite of all this, I enjoyed my duty at Philadelphia Naval Hospital. While there, I grew to appreciate the Navy, and was glad to have been classified a pekker chekker instead of an engineman.

CHAPTER 5

Continuous Fears

Fear is the foundation of safety

- Tertullian

There are many more stories from my first duty station, but gathering war clouds in Southeast Asia cut short my scheduled normal tour in Philadelphia. This resulted in me being transferred to the 1st Marine Division at Camp Pendleton, California, in 1960, two years before my tour in Philadelphia was set to end.

The 1st Marine Division at Camp Pendleton, the 3rd Marine Division on Okinawa and the 4th Marine Brigade in Hawaii were boosting their readiness and girding up for conflict in Laos, Cambodia and Viet Nam. I was not interested in politics and had no idea what was taking place. I just knew I had been cut short of my tour and transferred out west from a very fine duty station in Philadelphia.

Unknown to me at the time, President Kennedy was involved in serious negotiations regarding Prince Sauovana Phuma allowing the North Vietnamese to cross the Mekong River from Laos into

South Viet Nam. Though temporary actions were being taken to prevent that, it happened anyway, and was a major factor in the Ho Chi Minh trail and the reason why the North Vietnamese ended up victorious over South Vietnam many years later.

****

In September 1960 I was granted fifty days to get to Camp Pendleton in California. The Navy granted me thirty days leave, twelve days travel time and eight days of proceed time. The leave time was paid at a certain rate per day whether you took it or not. If you did not take the allotted leave time, you could sell the unused days back to the government. The travel time was paid at a specific mileage rate and could be taken before departing for your old destination or after you reached the new one. Proceed time was paid on an entirely different scale and was lost if you did not use up the allotted days.

I checked out of Philly with my travel orders and travel pay and headed home to North Carolina on leave. I was determined to use up all the leave, proceed, and travel time except what was needed for the train schedule to get me to California. I did all I could to avoid being AWOL and the court marital or ramifications of it. Now, there was a considerable difference between leave and precede pay, so I was determined to use it all so as not to lose any pay.

After using up all of my money, thirty days leave, eight days proceed and most of my travel time in North Carolina, I boarded a train in Charlotte bound for Camp Pendleton, California, with just four days left to get there. Being totally broke, I was eager to get back into the military routine and draw my back pay, which was accumulating nicely because I didn't know I could have drawn advance pay before leaving Philadelphia.

Before I left for California, my two oldest sisters, Johnsie and Fannie, prepared me a box of food to take on the train. The box consisted totally of crackers and canned snacks, but they didn't include a can opener. By the time the train was heading west out of New Orleans, I was starving and eager to get into my food but was unable to open the cans.

An elderly white man and woman boarded the train in New Orleans and sat on a seat facing me. They were continually munching on sandwiches from a big basket that was placed between them. I am sure they must have noticed the hunger in my eyes because they offered me some chicken. I eagerly accepted and when they discovered I was in the Navy, they nearly talked me to death. They shared their big basket of food with me and told endless stories about their son, who was in the Air Force, all the way to San Antonio, Texas, where they left the train. The food was good and very much appreciated. I regret the layover in Texas was not long enough for me to visit The Alamo.

As the train was leaving San Antonio, I asked the porter to take a can to the kitchen car and open it for me. Instead, he brought me a small can opener and told me to keep it. I was thrilled and snacked on my canned goods and crackers all the way to Oceanside, California, where the train pulled in late one night in early October of 1960. I didn't have a penny in my pocket and had no idea how I would get to the base.

I slipped back into the train station and changed into my uniform, intending to hitch a ride to the base. Grabbing my sea bag, I was able to get some Marines to give me a ride to the main base receiving area for Field Medical Service School. I arrived at Marine Corps Base, Camp Pendleton, California, around midnight in October 1960, to check in for Field Medical Service School training.

I was still about a month away from turning twenty-one and was not allowed inside any of the clubs near the train station. About a month later, I made it my business to return to that club on November 1, when I was allowed in at midnight to celebrate my 21st birthday. The bouncer remembered me, we had a good laugh and I had my first drink in California on the house. Back in Philly, the drinking age had been 18.

****

That training turned out to be exactly the type of duty I had joined the Navy to avoid. This was harsh infantry-type training, pure and simple. We tramped through brush, slept in fox holes in all types of weather, and sweated out rattlesnakes, scorpions, giant tarantulas and God knows what other creepy crawling critters that might have tried to get into our sleeping bags. Did I ever miss the clean sheets, hot showers, and steak and lobster meals of Philadelphia Naval Hospital!

During our five days of night training in Horseshoe Canyon I saw my first UFO, but I didn't think of it that way. We all thought of it as Sputnik and called it that. Everyone in the platoon witnessed it but no one mentioned the word UFO.

It was around midnight and we had just returned to base camp from a night-time map- reading exercise when some guy shouted out, "There goes Sputnik!" We looked where he was pointing toward the eastern horizon and observed a fast-moving multicolored light moving across the night sky. It reached the western horizon in just a matter of a few seconds, turned around and came back directly overhead. The captain said, "That ain't no damn Sputnik! Sputnik couldn't turn around at the horizon and come back here over us and stop like that!"

While he was talking, the object started to descend straight down into the canyon where we were camped and hovered for a few seconds overhead, shining a bright light on our campsite. The captain said, "It must be some kind of test vehicle." Just as he said that, the lights on the thing changed colors, and it zoomed straight up and out of sight in less than two seconds. We talked about that strange object all night, but nobody referred to it as a flying saucer or UFO. It would be at least thirteen years before "flying saucer" and "UFO" would become a part of my vocabulary.

That occurred when I witnessed weird flying objects in Norfolk, Virginia, along with several friends of mine. They were Albert Kelly, his wife Ronnie, Ronnie's mother Doreatha Medley, and her friend Lois Parker. We all observed five objects for several minutes, before a jet came over and chased them away. That sighting made the newspaper the next day as a UFO sighting and I've been pretty much a believer ever since. It was Labor Day 1975. I've been a UFO enthusiast ever since and have seen many other unusual object in the skies during subsequent years.

That evening as we sat on the porch there was not a cloud in the sky anywhere. Suddenly, we noticed four cigar-shaped dark objects approach the moon and stop equal distances from it at twelve, three six and nine o'clock. As they hovered motionless, a large black object appeared and blacked out the four objects and the moon. There still was not another cloud to be seen anywhere. As we watched and discussed this phenomenon, two jet aircraft approached from the direction of Hampton, Virginia. As they neared the objects, obscuring the moon, the larger object sped straight up and out of sight. The four smaller, cigar-shaped objects all sped away in different directions and disappeared except for one, which sped away across the Chesapeake Bay toward Maryland with both jets in pursuit but unable to close the distance before it disappeared. The two jets returned and made several passes over the Norfolk Naval Station before breaking off and heading back toward Hampton.

The newspapers the next day was filled with articles concerning the sightings the night before by numerous observers. This incident caused me to think back to the episode in Horseshoe Canyon during field medical training, when we all had referred to it as a Sputnik sighting. Sputnik no longer was a part of aerial phenomenon references. UFO and flying saucers had become a common phrase for describing unknown objects in the sky. To this day, I am a serious believer in unexplained things having the potential for being outside of the realm of human technology and capabilities.

****

After completion of the Field Medical Training I was assigned to the legendary 5th Marine Regiment of Belleau Woods' fame. I got attached to India Company of the 3rd Battalion as their only medic, where I became known as "Doc," the term of endearment and respect bestowed by Marines upon Navy medics assigned to serve with them. As part of the Navy, the Marines had no doctors, nurses or medics of their own. They got all their medical support and most of their other logistical support from the Navy.

With the 5th Marines, the tough infantry-type training continued in an even more stringent and intense mode than what we experienced at Field Medical School. During Field Medical training, we were all Navy corpsmen and could pretty much hold our own with each other. With the Marines, we were put into all types of simulated combat scenarios with young rugged Marines and real battle-hardened veterans from Korea and World War II.

The gung-ho, Marine recruits were mostly fresh out of boot camp and the Marine Corps advanced infantry combat training courses.

Along with those personnel were many kinds of Marines specializing in mortars, machine guns, rocket launchers, flame throwers, explosives, 106 recoilless rifles, reconnaissance, communications and other top-secret specialties. Whew! I wasn't even ready for all that necessary activity for war fighting!

How I wished I were still back in Philadelphia with clean sheets, hot showers and hot food every day. Especially those fantastic midnight rations prepared by Elmer, the night cook. I really missed being able to hang out at the arcade on Broad and Market, getting a facial, haircut, catching a movie on South Street and being able to stroll leisurely back to base watching all the pretty girls walking along Broad Street.

The entire time I was assigned to the 5th Marine Regiment I never saw another black medical corpsman. There were plenty young black Marines, including David Wiley, who lived near Mercer Street where four of my sisters and two of my brothers were living in Jersey City, New Jersey.

****

I met a young white man in the regiment who was also a corpsman. He was assigned to Kilo Company, next door to my company, and we formed a close friendship. He told me his name was Robert Jon Vanderlan and that he was from Utica, New York. We remained fast friends for the duration of our attachment to the 1st Marine Division. There were no other black medics assigned to our regiment. This became my first real social interaction with my white counterparts. Prior to this time my only real interaction with white youths of my age was limited to working with them on the farms back in North Carolina. Schools were still segregated in North Carolina when I left home to join the Navy.

Bob Vanderlan and I shared a lot of good times together on and off the base. We also deployed to Okinawa together for five or six weeks of temporary duty during the Laotian crisis in early 1961.

****

Returning from Tijuana, Mexico, around 2 a.m. one Sunday morning in early 1961 we ran out of gas and left our car about a mile from the barracks. We planned to go back for it later. As we reached our barracks we found our lockers cut, our bags packed and messages to report immediately to battalion headquarters with our bags. We did so, and found orders transferring us for Marine Corps matters beyond the seas. We were not told where we were going. We were given a series of several inoculations and whisked away to the helicopter pad to join several other medics who were waiting and also wondering what the hell was going on. When we returned to California several weeks later we were able to read in the papers all about the Laotian crisis and where we had been, and why.

We left from Travis Air Force Base on a C-130 cargo flight to Okinawa. The flight was mostly uneventful except for a stop in Hawaii to repair an engine that was leaking oil and a refueling stop on the tiny Wake Island, which was still littered with visible wreckage from the battles fought there during World War ll. I stayed on Okinawa for the duration of the emergency temporary deployment. Bob got sent into Laos with a Marine Expeditionary Brigade. I was moved around from base to base-on Okinawa and was afraid I might never get back together with him again.

We never saw our abandoned car again. The clothes we had laundered before leaving for Tijuana that Saturday and those left hanging out to dry were never seen again. Incidentally, we never got reimbursed for those clothes either.

****

When the crisis was over Bob and I were reunited back at the 5th Marine Regiment in California. We renewed our association, and spent time swapping tales about our adventures on the emergency deployment. Bob had been sent with a unit of Marines to secure the air base in Laos and was photographed for the cover of Life magazine riding a water buffalo. I did absolutely nothing except sightsee and enjoy Okinawa from Naha at the southern end to Camp Schwab at the northern tip of the island. I enjoyed attending the snake and mongoose fights. We named all the mongooses Rikki Tikki Tavi. They were really swift little creatures and very seldom lost to the snake. It was a painful sight when they did loose.

On Easter Sunday on Okinawa 1961 a lot of us got recruited to play extras during the filming of the movie "Marines Let's Go!" Had I known how they were going to use us, I certainly would not have volunteered. They ran us up and down a steep hill seven or eight times, gave us two baloney sandwiches, some lemonade, and sent us on our way. We got all of fifteen seconds of on-screen time in the finished product and the only thing that made us recognizable was the big red cross on the ambulance used in the assault on the hill.

The bulk of the Marines and soldiers from Okinawa and Hawaii were deployed in support of the Laotian crisis, which left little for the skeleton crew on the island to do. I did get to see what a monsoon was like when we got stranded in the enlisted club for two whole days and nights at Camp Schwab. But mainly I just spent time exploring the island and old World War II ruins. I visited the legendary Teahouse of the August Moon at Naha, and subsequently named one of my daughters after one of the cute Geisha Girls I saw there.

****

When we returned to California, Bob went back to "K" Company. I returned to Company "I", my old unit, which was quartered next door to his. We were thrown into an intensive training regimen, which included cold weather and desert training. During one live-fire exercise, I witnessed my first Marine death when an unexploded mortar round killed a Marine as they swept the firing range for duds after the live firing exercise was completed. We medics had been kept way back from the range on a hill, but we rushed to the scene when the range monitor yelled out, "Cease fire! Actual casualty!" When we got to the injured Marine, there were only scraps of him to recover. That was my first taste of what real war could be like. I will never forget it even though, as a medic, I saw many more casualties during my Navy career.

Again, Bob and I became inseparable and spent many leisure hours together. We rented a 1951 Chevrolet together and spent time visiting Los Angeles, San Diego, Oceanside and Tijuana, Mexico.

****

One weekend in Los Angeles, David Wiley and I got tossed out of a club by the cops we had called. When the club manager refused to give us change back after we bought drinks, we summoned the police. The cops clubbed David in the back with a baseball bat the bartender had leaning against the bar. This was done as he tried to explain that the three white Marines at the bar had invited us into the club for drinks.

The cops threatened to lock us up and then ordered us to leave the club without our change. It was Marine Corps Birthday Ball weekend of 1961 and we were in our uniforms. I never did get back my switchblade knife, which I had hidden behind the door before the cops arrived.

Hanging with Bob was enlightening. He schooled me on many overt and covert racial attitudes and practices of the white guys in our regiment. Through him I learned of many racist jokes and derogatory names for blacks popular around the base. Names like "spear chucker," "night fighter," "coons," and "jungle bunnies" were used to describe us.

We got discharged at the same time in January 1962 at Camp Pendleton, California. After being processed out, we pooled our resources and bought that old 1951 Chevrolet we had rented and drove it cross country to New York with only a couple of minor incidents.

****

The first incident happened in Yuma, Arizona. While I was filling the tank with gas, Bob went inside and ordered us a couple of breakfasts. When I tried to join him at the lunch counter, I was not allowed to sit and eat with him. My plate was taken back into the kitchen. Bob, threw his plate on the floor as we ran out and made a hasty departure from those premises. Bob was so upset; he threatens to toss a hand grenade we had through the back window. Somehow, we found our way back onto the main highway and had no more trouble until we reached Missouri. There, the alternator went out and we had sense enough to let Bob negotiate the repairs, which turned out to be reasonable.

From there, we made it to Grand Central Station in New York without further incident except for a brief snowstorm in Pittsburgh. We were a bit weary from driving virtually non-stop all the way from California, but elated to be back on the East Coast.

I bought out Bob's interest in the Chevy at Grand Central Station. Promising to keep in touch, we said our goodbyes and went our separate ways. I never saw Robert Jon Vanderlan again. I tried many times over the years to find him but was only able to connect with him by phone a couple of times. After he retired from Sears in Utica, he moved his family to Florida. He was a good guy; a really terrific friend and I have never encountered the likes of him since. I wish I had known Max Moore, Dewey Moore's son, in the post-civil rights era. I bet he would have been a terrific friend just like Bob Vanderlan.

****

As Bob left Grand Central for Utica, I headed across the river for Jersey City, New Jersey, to visit with some of my sisters and brothers. I spent a few wonderful days in Jersey with them before heading south to North Carolina to visit with my father, mother, baby sister and baby brother, who were still helping my father sharecrop. I spent about three weeks and all my money at home partying with Gene Thomas and Herbert Louis Tillman. During this time I found out why they did not want to go in the Navy. I headed back to New Jersey to look for work thereafter. By now you are probably getting the picture that I was spending money like a drunken sailor. The truth is that I was seldom drunk.

Back in Jersey, I was lucky enough to get re-hired by Gold-Pak meatpackers in Union City, where I had worked my first summer out of high school. More mature now and somewhat more liberated, I was not willing to put up with the unfair and unsafe working assignments I had accepted four years earlier. So after a couple of weeks, I went to Brooklyn in search of a Navy recruiter to see what kind of deal I could wrangle if I re-enlisted.

****

I would never have left the Navy if I could have been transferred to the East Coast from California when Bob and I were discharged. The Navy refused, so I took my honorable discharge and headed east to New York with Bob Vanderlan. I wish Bob could have been there with me when the unlit cigarette incident occurred later at Camp Lejeune. I'm sure my entire military career would have turned out quite differently and much better.

When I re-enlisted, I'd had enough of the Fleet Marine Force and hoped I would finally get a shipboard assignment on the East Coast. That would not happen for several more years.

In March of 1962 I was headed right back to the halls of Montezuma and the shores of Tripoli. Literally. My re-enlistment netted me an assignment to an aircraft carrier stationed in California. As luck would have it, another corpsman re-enlisting the same day was given orders to the 2nd Marine Division based at Camp Lejeune, North Carolina, and wanted to swap orders with me.

His home was on the West Coast; mine was on the East Coast. The recruiter agreed and we switched. I accepted his orders and got a couple thousand dollars pay upon re-enlisting. This was quite a nice sum of money for a completely broke young man at that time.

After being out of the navy for only fifty-six days, I was right back in and headed to an assignment with the Fleet Marine Force on the East Coast. This time I was being assigned in my home state and much closer to home.

CHAPTER 6

Final Fears

"The first and great commandment is, don't let them scare you."

- Elmer Davis

When I reported to the Second Marines I got assigned to Charlie Company of the 1st Battalion, Second Marine Regiment, as the senior company medical aid man. After six months of rigorous training, constantly sweating cottonmouth moccasins, copperhead snakes, rattlesnakes, red bugs, ticks, and sometimes alligators, we shipped out to the Mediterranean in September as a floating battalion with a Marine Expeditionary Brigade.

The only good thing about the pre-deployment training was the frequent trips I could take home to Peachland, North Carolina. An added benefit of being a navy corpsman attached to the Marines at that time was you could choose to wear either your Navy or your Marine uniform home on weekend passes. I had fun confusing the heck out of the people who saw me in a different uniform every two or three weeks. They could not believe I was actually in the military. I got a lot of laughs out of that situation while also thoroughly impressing the hometown girls. Other than the respect the Marines accorded us "docs," serving in the field with a combat unit, certainly was not my cup of tea.

This deployment would be my first of three to Europe. It was fairly interesting though I mainly just saw the insides of clubs and bars when off duty. The cruise was subsequently extended to seven months due to the Cuban Missile Crisis, which happened while we were on training exercises in Tripoli and Bombay, Libya. After the end of the crisis we moved to Corsica for more exercises.

On November 16, 1962, around midnight, while we were on a training exercise on the island of Corsica, I was summoned to company headquarters located about a mile from my platoon. This was a really frightening experience tramping alone through the brushy terrain in the dark, but I figured somebody at headquarters was in need of some medical attention. I was still deathly afraid of snakes and had seen several black and gold ones among the rocks and brush during that day. I arrived unscathed at the company command post and was greeted at the campfire by Capt. J.R. Prochasko, who was accompanied by 1st Sgt. Robert W. Carter and the radioman.

Prochasko, the company commander, greeted me with "Congratulations, doc. As of midnight tonight, you were promoted to Hospital Corpsman Second Class. You are also selected to attend X-ray technician school at Bethesda, Maryland, providing we get back to the states in time. Your class starts in March 1963." I was overjoyed and floated on air over all those snakes as I returned to my position with the weapons platoon.

I was the only medic in the battalion to be promoted to E-5 that November. Thanks to President Kennedy's resolution of the missile crisis, a relief battalion got to the Mediterranean in time for us to return to Camp Lejeune two days before my training was scheduled to start at Bethesda, Maryland.

****

I said goodbye to field medical duty for the last time and completed the six months of classroom training at X-ray school in Bethesda. Upon graduation, I received orders to return to the Marine base at Camp Lejeune Naval Hospital for six months of on-the-job training. My departure for Camp Lejeune was delayed a few days due to everybody being confined to base because the 1963 march on Washington led by the Rev. Martin Luther King Jr. was in progress.

Back at Camp Lejeune my training required me to rotate through the various duties of the X-ray department. In October, while I was working in the darkroom with the radio on, I heard that President Kennedy had been shot. I rushed out into the waiting room to tell everybody. A young eye-clinic technician named Lopez, who worked next door, clapped his hands with glee and shouted, "Great! I hope he dies." Outranking him, I reprimanded him for his display of disrespect for the President. Sometime later, he would be in charge of the barracks cleaning detail and place me on report for failing to carry out my cleaning assignment.

Once again I was charged, convicted and punished. Even though I outranked him, I was assigned as a student and placed under him for cleaning details. Not long after that, in December, the unlit cigarette incident occurred at the field house and my military career would forever be derailed. Never again would it be the same for me in search of promotions. I had been in the Navy for six years by this time. I had succeeded in getting five promotions in that time. I would serve fifteen more years but only get two more promotions. Three, if you include the one I would lose at courts-martial but gain back.

The unlit cigarette incident was definitely the straw that broke the back of higher promotions for me. I certainly didn't know it at the time, or I would never have made the Navy a career. I was used to overcoming all my setbacks by moving on up the promotions ladder.

****

After completing my tour of school duty at Camp Lejeune, I was transferred to Preventive Medicine Unit 2 in Norfolk, Virginia, to work in the public health department tracking and monitoring tuberculosis. Our three-man X-ray bus crew consisted of James H. Bebout, Rudolph V. Nickels and me. It was a good assignment and I really enjoyed it. While on this assignment, I fought hard to overcome the negatives in my record acquired at Camp Lejeune. I earned promotion to Hospital Corpsman First Class (E-6), a nice step up the enlisted promotion ladder for a young medic not far removed from the cotton fields of North Carolina. While enjoying this tour of shore duty, I married Vermel Polk, an old friend from back home whom I adored. I worked many part-time jobs during this period to send the apple of my eye to Norfolk State College. I managed the pool hall on base, operated the movie projection booth on base, ran the snack bar at the enlisted club on weekends, made pizzas at a grocery store and loaded trucks at night for Seal Test ice cream.

In October of 1966, I was ordered to Iceland for ten days to administer flu shots in place of another medic assigned to disease and vector control. While I was in Iceland, the Red Cross notified me of the birth of my daughter. I was supposed to

88

have been present in Charlotte, North Carolina, for her delivery. Tomiko Carmel was born October 19, 1966, while I was still in Iceland doing someone else's job. My absence from her birth and my frequent deployments might have contributed to the disdain my wife subsequently felt for the military. She subsequently developed a career of her own with Eastern Airlines and filed for divorce while I was still in Vietnam.

None of the three automatic flu shot guns worked in Iceland and I could not complete the mission. I found out years later that I had been sent on an experimental run to evaluate whether the newly acquired flu guns would work in that environment, while the corpsman who was to do so was allowed to go on leave to his home in California.

The new guns didn't work at all and the shots were administered the old fashion way with needle and syringe by the corpsmen already stationed at Keflavik. The corpsman that I replaced on the Iceland trip was well aware of the major malfunctions inherent in the new inoculation guns, as he explained to me upon his return from California.

****

In 1967 my shore duty again was cut short because of the conflict in Southeast Asia, which had become a full-fledged raging war. I received orders notifying me of my transfer to a Minesweeper, the USS Notable MSO-460, at Charleston, South Carolina. I had completed only one half of my scheduled tour in Norfolk at Preventive Medicine Unit #2. The promotion I received in Norfolk was the next to last promotion I would ever receive during the remaining 12 years of my service in the Navy career thanks to an unlit cigarette.

While serving in Charleston, I was switched to another Minesweeper, the USS Ability MSO-519. Then the four Minesweepers of our squadron were sent on a six-month Mediterranean cruise, my second.

Our Mine Sweeping Squadron had participated earlier in the recovery attempt of the nuclear bomb lost off the coast of Spain. That was the operation where the first black master diver lost his leg saving several lives during the recovery operation. He later got an artificial leg and was restored to full diving status. Bill Cosby would later produce a movie titled "Men of Honor" starring Cuba Gooding, Jr. depicting the master divers life story, known to us today as Carl Brashear.

The USS Rival and the USS Salute accompanied the Notable and the Ability. I was moved to the Ability because it was the senior ship in the squadron and carried the commodore. I was the senior medic assigned to our squadron and authorized to perform the duties of a medical officer ashore and afloat.

Each minesweeper had its own captain and the commodore commanded the four captains. While serving aboard the Ability, I continued to overcome bad evaluations and collected the last promotion I would receive during my Navy career. Where promotion was solely dependent upon proficiency, ability to pass an exam, and do the job, I always excelled.

****

While serving in Charleston, I made many weekend trips with a friend, Albert Kelly, to Charlotte, North Carolina, to visit our families. During one of those trips, we came upon a Ku Klux Klan roadside bazaar along the side of Highway 52 near Florence, South Carolina. There were all kinds of paraphernalia on sale so

I bought some 45-rpm records, including "Flight of the NAACP" and "Move Them Niggers North." I thought they were quite creative in their racist promotion. I was extremely disappointed when my younger brother snuck those records onto his base at Fort Bragg and got them confiscated. I would love to have them today to show my children and grandchildren a picture of another place and time in race relations.

****

Despite all of the petty disciplinary actions accumulated in my record, my performance overshadowed them and I still gained my final promotion, effective July 16, 1969, to the coveted position of chief petty officer (HMC-E7), known as "getting the hat," while still onboard the USS Ability-MSO 519. I would be on my way to Vietnam before I was authorized to wear the chief's uniform. After we returned from the Mediterranean cruise, my tour was cut short again and I was ordered to proceed from the USS Ability to report to Viet Nam.

HMC-E7 was the highest enlisted rank one could attain without the input of a board review. It was my last promotion and I still had ten years to go before I would retire. In ten years I secured all seven of my promotions. Through the next eleven years, my record showed, despite the many disciplinary actions against me, that I was both highly qualified and intensely recommended for higher promotions as well as officer selection but I would not get either.

While deployed to the Mediterranean aboard the USS Ability, I had to manage several medical conditions including an appendicitis attack, a minor heart attack, a case of meningitis and many cuts and bruises resulting from bar brawls or training accidents.

****

One night while anchored at Palma, Spain, after attending the bullfights, me and my friend, Thomas V. Peterson, stopped off at a pub on the way back to the ship and got wound a little too tight. After we were sufficiently oiled up at the pub, we convinced ourselves to go aboard The Ability to the Captains stateroom and continue a discussion with him on race relations, which we had started at a picnic a few days before.

We knocked on the Captain's stateroom door and were invited in to find the Captain's cook serving snacks to Zsa Zsa Gabor, Peggy Lipton and another fellow. The duty medic was applying a bandage to Zsa Zsa's legs where the Spanish police had injured her while trying to remove her from an airplane. She had a little basket with a tiny dog in it. The basket had a false bottom that contained, I swear, at least a million dollars in jewelry. They stayed in the Captain's stateroom for a couple of days until arrangements with the embassy could be made to get them out of Spain and back to America. Peterson and I never finished our planned confrontation with the Skipper. We quickly excused ourselves and made a hasty retreat. Before her entourage departed, Zsa Zsa gave the captain's cook, Crawford, whom we called "Crow," a very expensive looking bejeweled cross. I was supposed to have been the medic on duty but had traded off so I could go to the bullfights with Peterson. I never found out if she gave the duty medic anything.

Shortly after arriving back in the states, Peterson and I got promoted to chief petty officer. He was the ship's storekeeper and I was the medic. I received orders to Viet Nam. I don't know where Peterson got orders to, but I would eventually run into him managing the chief's club at Camp Tien Sha in Danang. Once again I was cut short of a scheduled tour I was thoroughly enjoying at my Charleston duty station.

I could have deferred my transfer to Viet Nam at this particular time because my baby brother was serving there with an Airborne Brigade. I could have deferred until he rotated back stateside but the detailer in Washington, D.C., assured me if I went now he would guarantee me an assignment on a hospital ship to work in X-ray. So I took those orders so as not to be assigned to a combat unit with the Marines, which was the assignment detailed in my orders.

My detailer was as good as his word. But had I known what other horrible experiences awaited me in Viet Nam I would certainly not have gone at that time.

****

Back when I had received my orders to leave PMU2 and report to Charleston I had re-enlisted and used the money to purchase my first home in Charlotte, North Carolina. That was so I could leave my wife and baby there near her parents, while I got settled aboard the USS Notable MSO-460. It was a nice house and a very good deal. The white folks who owned the property were getting divorced, eager to sell and get out of the Clanton Park area, which was rapidly turning into a predominantly black neighborhood.

The husband was career military with orders for Germany. The wife was managing The Ramsey Lewis Trio and preparing to go on tour. They let me have the house and considerable equity in it, for just $900 and loan assumption, a really low price for the fairly new home, which had a large lot.

While I was stationed at Charleston, my youngest daughter, Victoria Kymyarda, was born. She celebrated her first birthday the month I left for duty in Viet Nam. I will always remember her crawling all over me while I was trying to watch the first landing on the moon while home on leave. Tomiko, who was three years old, was sound asleep and my wife was at work with Eastern Airlines. Kym was always such good company and I loved to see her come dancing down the stairs whenever I played Miriam Makeba records. She especially loved it when Miriam Makeba was singing "Nomtheni."

****

Standing on the tarmac at Douglas International Airport in Charlotte on that July day in 1969, waiting to board an Eastern Airlines flight, many thoughts rushed through my mind as I listened to country singer Johnny Wright crooning "Kiss me goodbye and write me while I'm gone. Goodbye, my sweetheart. Hello, Viet Nam." I was standing there with my wife Vermel, holding little Victoria in my arms. Tomiko clung to my leg, screaming for me to pick her up as the jet engines revved up. My heart was torn apart and I ached to pick her up, but refused to do so as I thought about the reality that she might never see me again.

I handed Victoria to her mother, kissed them all goodbye, and boarded my flight for the war zone with much trepidation. Not picking Tomiko up in that moment was about the hardest thing I've ever had to do.

****

After a two-day layover on standby in Atlanta and a three-day stop in California to visit my friend, Albert Kelly, and his wife Ronnie, I boarded a flight to Viet Nam to report to the U.S. Naval Hospital ship Sanctuary AH-17, which was anchored about two miles off the coast of Danang. After a very apprehensive one-night stay at Camp Tien Sha near Monkey Mountain, we were given a fast skimmer boat ride out to where the ship was anchored.

After landing at the Danang Air Base the night before under the cover of darkness, we were treated to a harrowing ride through narrow streets of the dark town in a truck screened with wire to protect us from hand grenades while artillery put on a very awesome display of sight and sounds on the horizon. As I watched the night sky being lit up by bombs and artillery fire, we snaked our way through the darkened streets in the wire-caged truck. I said a silent prayer that I would make it back to my family safe and sound as I ducked low in the wire covered truck. The truck ride was followed by a sleepless night at Camp Tien Sha to await the skimmer boat ride out to the Sanctuary the next day. The skimmer traveled at a very high rate of speed to make it difficult for Viet Cong rockets to zero in on us from Monkey Mountain.

The hospital ship off the coast of Viet Nam was a welcome sight. After the long plane ride from the states, and a scary truck ride through darkened narrow streets, it hit me that this was a real war zone. It quickly dawned on me that this was a place where death and destruction lurked for real and a farm boy from North Carolina could get killed. It was hard to accept the fact that this was going to be a way of life for me for the next year. Welcome to a world of real fear and destruction!

****

My promotion to chief petty officer became official on July 16, 1969, while I was en route to Viet Nam. A couple of days after I boarded the ship, it left Viet Nam for a routine ten-day trip to Subic Bay in the Philippines to pick up supplies and offload patients. Traditionally, we made this trip every three months. After my fourth trip to the Subic Bay in a year's time I was scheduled to depart for the states from Viet Nam.

I was assigned a bunk in the chief's quarters and hung my trousers on my bunk. I went into the restroom to freshen up while a young sailor from deck division was cleaning the quarters. After a few minutes I came out of the washroom and found my wallet and the young compartment cleaner gone.

I reported it at once to the command chief petty officer, which also happened to be in charge of the person assigned to clean the quarters. He responded by berating me for accusing his man and asking that his locker be searched. He also refused to have his man questioned or searched even though I outranked his man by four pay grades. I never got my wallet or its contents back. I was in the washroom for less than five minutes while the young sailor was in the compartment alone. This was a very disheartening lack of respect and disappointing welcome aboard for me as a new chief petty officer. This set the tone for the rest of my tour in Viet Nam.

Just a few months later, another of the command chief's men, whom I also outranked by four pay grades, charged me with creating a disturbance in a boat. This accusation resulted in the command chief petty officer charging me with assault and disobeying orders to retire to my quarters.

****

I had been assigned to the hospital ship to supervise the X-ray department, but instead was detailed to supervise the most difficult division in the hospital, the 132-man patient handling division. This was an immediate giant leap from supervising a crew of one, myself, to supervising an entire division of men. That division was under the direction of Chief Nurse Ann O. Watson. She exhibited total disdain for me as evidenced by giving me the worst evaluations I ever received as a chief petty officer, including poor appearance reviews, though I was one of the neatest appearing chiefs on the ship. I was trim and fit, dressed in my brand new uniforms.

I was the most junior chief on the ship's hospital roster so when my brother rotated back to the states a request came in for an independent duty-qualified person in the Delta. I was immediately removed from the hospital ship and dispatched to the USS Clarke County LST-601 operating out of Vung Tau in the Delta because my brother had unexpectedly been sent stateside.

A request came in for an independent duty-qualified person in the Delta. My brother was unexpectedly rotated back to the states. Since I was the most junior chief on the ship's hospital roster, I was immediately removed from the hospital ship and dispatched to the USS Clarke County LST-601 operating out of Vung Tau.

****

The Clarke County was one of the many flat-bottomed ships operating throughout the Delta Rivers, waterways and tributaries. It was tasked with delivering supplies to the many barracks ships, repair and maintenance ships. It also provided the many repair, berthing and messing barges of the Riverine forces with mobile support. These were the same Riverine Forces of which former presidential candidate John Kerry had operated while he was in Viet Nam.

I got aboard Christmas Eve in Saigon and we headed south to Vung Tau along a narrow river under a bright moon. There was only one other chief assigned to the small crew. He was an engineman who spent most of his time in the engine room making water and steam. For the first few hours he spent time with me in the chief's quarters, orienting me to my responsibilities and my general quarter's station, which was in the officer's stateroom.

After a while he left me alone watching movies while he went up on the bridge to talk with the captain. Suddenly, there was a loud explosion, which caused the ship to shake violently. In an instant, I was out on deck and the engineman chief shoved me back inside and said, "Doc, stay below and keep the ship dark. That was just our guys throwing concussion grenades to keep the V.C. swimmers from planting explosives on our hull." That was not very reassuring since the moon had disappeared and we still had several hours to go before reaching our base at Vung Tau. That was the damndest Christmas Eve I ever spent anywhere!

There were many scary and dangerous incidents while I was aboard the Clarke County in the Mekong Delta. Once we got stuck on a sandbar at the mouth of two rivers and remained stranded there for several hours. A ship like ours, the USS Lazerne County, had been struck with a rocket with a large number of casualties, including fatalities.

We waited for the tide to come in to lift us so our anchor could drag us off. Meanwhile, Swift Boats and PBRs (Patrol Boats, River) circled us, constantly firing into the jungles around us the entire time. This was to prevent us from being blown out of the water and losing our heavy load of supplies, which had contributed to our getting stuck there in the first place.

I really felt bad for the young black lieutenant junior grade who was controlling the ship when it got stuck. It really didn't matter who was at the helm. Even if it had been the most experienced officer on the ship, we still would have gotten stranded. The sandbar wasn't there four days earlier when we navigated the area empty. We were now fully loaded and the sandbar was newly formed and uncharted.

Another time, a Filipino barge behind us, and a Cambodian barge in front of us, came under furious attack; forcing us to go to general quarters and give covering fire. With our Swift Boats and PBR escorts, we ran the gauntlet of V.C. assaults from both sides of the narrow river. Being locked down at battle station with no way of knowing what's going on, waiting for the first shell to slam into your ship is about the worst way I can think of to take a South Pacific cruise.

Every week one of our LSTs delivered food, fuel, ammunition and other supplies to the river forces. From Dong Tam and Cat Lo to Bien Thuy and many other outposts, supplies such as barbed wire, beer, soda, clothing, foodstuff, medicine and bags of concrete mix were distributed each trip.

The LSTs were constructed in such a way that, when necessary, could drop their anchor and run their bow on to the beach to be unloaded either by hand or lift trucks. Watching the Vietnamese clean out the ship after unloading was a sight to see. There would often be spillages. They would clean everything up and keep the waste as compensation.

Their favorite salvage item was the spilled concrete dust. For some reason they valued it over the beer, soda and cigarette wastes. They would argue over who would be allowed to clean it up. When they finished cleaning the ship's well deck, there would not be a speck of concrete dust left. I have yet to figure out why concrete dust was so highly prized by them. At each delivery point, I would treat as many minor medical conditions as my medical supplies would cover. Skin rashes, parasites and venereal disease were the most frequently encountered complaints.

After a few weeks, the guy I had been sent to the Clarke County to relieve returned from the states. We knew he was back but could not find him. Eventually we tracked him down in a brothel, where he was totally drunk, and lugged him aboard his ship. I got him sobered up enough to accept the keys to sickbay and the combination to the narcotics safe. Only then was I able to convince the captain to let me return to my hospital ship, the USS Sanctuary.

Those six weeks in the Delta aboard the Clarke County were really a tense time for me because several ships in the Clarke County's class had gotten rocketed by the Viet Cong, resulting in numerous casualties including many fatalities. I made it to the Halo-pad just in time to catch the last chopper heading to Saigon for the next three days. I reached Danang and returned to the Sanctuary just in time for a scheduled trip to Subic Bay.

I just missed the Bob Hope USO Show. However, I was mighty pleased to be back aboard and finally given the responsibility for the X-ray department that I'd been promised when I was first ordered to Viet Nam. I remained in that position for my last few months in Viet Nam until I rotated back to the states in August of 1970.

****

A few days after I reported to the USS Sanctuary at Danang in August of 1969, it left for a ten- day visit to Subic Bay in the Philippines. On our second visit to Subic, I was able to convince my brother's commanding officer at the Army's 80th Group to let me take him along for R&R. I had visited with my brother once at China Beach, where he was billeted when I first arrived in the country. While sitting around with a few of his buddies in their sand-bagged drinking pub, a nearby artillery unit decided to let go with an ear-splitting, earth-shuddering barrage. I looked around and saw I was the only one who had even moved. Everyone was splitting their sides with laughter.

One old grizzled sergeant said, "Chief that was just outgoing artillery. You don't have to worry about the ones you can hear. It's the ones already exploded, when you hear them that you have to worry about. And, really, you don't have to worry about them either because you will probably already be dead." That did little to allay my anxiety about being on the ground in a war zone. Even though China Beach was quite inviting, you can rest assured I stayed pretty close to the hospital ship from then on.

****

Upon completion of my Viet Nam tour of duty in August of 1970, the hospital ship Sanctuary set sail from Viet Nam to Subic Bay on its scheduled resupply and patient-transfer run. This trip to Subic Bay would be my last and I would disembark there to rotate back home to Charlotte, North Carolina.

As I boarded a bus in Subic Bay for the trip to Clark Air Force Base in Manila, a young man snatched my watch and disappeared down an alley. I would replace it with a nice Seiko watch when we stopped in Japan. That was the only expensive watch I ever owned.

The bus ride was a bit scary because thugs were constantly ambushing and robbing travelers between Subic and Manila. I thought how ironic it would be to complete my tour of duty in Viet Nam and get killed on a bus while returning home. Except for one security check along the way to Manila, the bus ride was uneventful. It was painfully slow, but uneventful. I was dog-tired because I had not slept a wink the night before because I was so anxious to get home.

After boarding the flight for home at Manila, thoughts of Viet Nam began fading into the rear view mirror of my mind. I had long ago determined if I were fortunate enough to make it through my tour in Viet Nam I would desert the Navy before ever returning there again. I was not a conscientious objector but was definitely anti-war, especially that war. While there, I had been treated very unjustly and with such lack of respect as a chief petty officer and as a man. That made it extremely difficult to appreciate the CPO Creed I'd worked so hard to earn and cherished so dearly.

As the plane taxied away from the terminal and sped down the runway, I wondered if the heaviness of my heart would be too much weight for the plane to lift off. I was elated at the prospect of departing Southeast Asia, but also very apprehensive about my return to the U.S. As the plane rose above the clouds, heading for our first stop in Japan, I was filled with some of the same feelings I'd had back in Marshville, North Carolina, when I first boarded that Trailways bus eleven years earlier to go enlist in Raleigh.

Through the plane's window, I filmed the green hills of Manila Bay with my Super 8 camera, watching as they faded into the distance. I was fearful about how I would deal with what was waiting for me at home. I had only received one letter from my wife the whole year I was gone.

****

The plane roared through the heavy cloud cover heading for our first stop in Japan. As the scenery below disappeared, home still seemed so far away as I dozed off.

Upon my return, I was assigned as X-ray department supervisor at Naval Hospital Newport, Rhode Island. The flight filled me with a mixture of joy and anguish. Joy that I was on my way home and soon to be reunited with my family again, anguish because I had only received only one letter the entire year I was in Viet Nam. That letter from my wife had been to inform me she had rented out our house, moved into her own home, and filed for a divorce. She said she could no longer put up with the frequent transfers and long separations and had a career of her own.

That flight felt like the longest of my life. We made stops in Japan, Alaska, San Francisco, Dallas, and Atlanta. In Dallas, some senator invited me to travel with him in first class and talked my ears off about the war all the way to Atlanta. I wish I could remember his name, but I was not paying much attention to him. I just soaked up his free drinks and continued wondering what awaited me at home. At the same time, I was trying to concentrate on an article in Saga magazine about a psychic town called Casadega in Florida. When the senator deplaned in Atlanta, I was able to finish the article on the last leg of my trip to Charlotte. Years later, in 1974, I would be reminded of that article in Saga with eye-popping and astonishing revelations.

My brother Bobby had visited with me in Danang at the Club White Elephant, while we were both in Viet Nam and played Tammy Wynette's song "Divorce" on the jukebox. I'd never heard it before, but he knew I liked country music. He had heard about my pending divorce back home and that was his way of trying to alert me to the bad news.

The plane had barely touched down at Charlotte Douglas International in the rain before I was on my way to my house on Clanton Road. As I walked around outside of the house, I could see the renters had moved out. The house was empty of all of my personal belongings, which had been ransacked and piled in the backyard, where they were soaking up the steady August rain. All the memorabilia and memories I had accumulated during the first eighteen years of my life and the first ten years in the Navy were either missing or ruined. Photo albums, prom pictures, stereo records, certificates of achievement, cruise books, my high school diploma and even the guitar that was given to me twenty-five years earlier were all gone or totally destroyed.

During my 42-day stay at home, the rest of my family and friends tried to cheer me up with parties, picnics and cookouts. Still, I don't believe I will ever recover from that heartbreaking homecoming from a war I so detested in the first place. The part that was especially depressing was when my 4-year-old daughter stood up in her chair and pointed out a man to me in Hoover's restaurant on Statesville Avenue and said, "Daddy, that's the man who took off all his clothes and got in the bed with Mama."

The man she pointed out was a neighbor in the Clanton Park neighborhood. He was married to a woman whom I had known quite well from the days when we both attended the Rev. Bowen's churches. She lived in Wadesboro and I lived in Peachland back when I was in high school, and we both sang in our respective church choirs.

She had confronted me once and explained how she had broken our living room window. When I arrived home from Charleston one weekend and asked my wife about the broken window, she said only that someone from down the street had broken it. I questioned people up and down the street and She admitted she had done it. She took me into my bedroom and pointed out blood spatters on the ceiling, which she said she caused when she struck her husband in the head with her shoe. She gave me her side of the story and we confronted my wife when she got home from work at Eastern Airlines that night.

I secretly recorded the conversation between the three of us and played it for my mother-in-law. She promptly pulled a gun and ordered me from her house.

When my daughter pointed the neighbor out to me, I went out to get my gun from the car only to find that I had left it at home. Thank God because I was military trained not to pull a gun or point it at anyone unless intended to shoot. I certainly would have shot the guy because my heart was breaking for my little girl having to endure such memories.

I let the matter drop and have long since forgiven my ex-wife. I hope my daughter has forgotten it too. As for me, I had long gotten over it along with the divorce. After I retired from the Navy, my ex-wife deserted the children and they came to live with me for a while. I never mentioned that restaurant incident to her or the children again. But I've often wondered if my daughter remembers and how it may have affected her.

My mother-in-law would not tell me where I could find my wife and children when I got home from Viet Nam, but my father-in law did. I was so very much looking forward to seeing my kids and not having to be away from them again for a long, long, time.

I now owned the coveted rank of chief petty officer and would have a measure of control over my future assignments. I had been looking forward to my wife participating with me in selecting our future assignments. After finally agreeing to her divorce terms, I reported for duty at Naval Hospital Newport, Rhode Island. I served one year of the scheduled three-year tour there before once again being sent back to Camp Lejeune, North Carolina, to manage the X-ray department at the naval hospital there. That would be my final tour of duty with the Fleet Marine Force at Camp Lejeune.

There were quite a few interesting occurrences during my year in Newport. I enrolled in some college courses at Salve Regina College during my off-duty time. The college was located in close proximity to the Bouvier Estate, the childhood home of Jackie Kennedy. I often passed near it on the way to and from classes.

Bill Cosby Sr. was hospitalized in Newport for a short while with a leg injury and the commanding officer had his picture taken with both father and son and hung it in the hospital lobby. That large picture was still hanging there when I left for Camp Lejeune.

I rented a small apartment off the base and my landlord, Joe Belcher, was a retired Navy man who had served with Bill Cosby Sr. They were long-time acquaintances and socialized frequently while he was a patient at the Navy hospital, where Joe, now retired, worked in the laundry.

One night I swapped duty with James Wilson as chief of the day. That night the fans rioted and ended the Newport Jazz Festival forever. I had tickets for the events scheduled for the next day and was looking forward to attending. Once again, it seems I got stiffed by swapping duty days. The show was completely turned out that night as Dianne Warwick sang "What the World Needs Now Is Love." I got lots of good footage of the festival's aftermath with my Super 8 camera the next day. I still look at the movies from time to time and wish I had gotten the opportunity to attend that historic final festival. The entire concert could be heard clearly from my duty location at the hospital but the entertainers could not be seen. I had tickets for the next day's concert and my father, sister Fannie, a brother-in-law, Clarence (Ditts) Thomas, and some close friends came up for a visit. But the festival was canceled so I took them on a tour of a large ship anchored in the harbor.

I had the good fortune of being able to commute to Jersey City on occasions to visit with the many relatives I now had living there. By far, the most amazing thing that occurred while I was at Newport involved my mother's younger sister, Aunt Mary Ross Tatum Smith. I had not seen Aunt Mary since I was five or six years old and living at Hamp Brewer's farm. Everyone had lost touch with her when she moved to Texas. For a while, she would send me birthday cards or a short letter.

A fellow sailor stationed at Newport invited me and a friend of mine, James Wilson, who was from Sanford, Florida, to a fish fry one Friday night in Providence, Rhode Island. When we walked in to the after-hours joint, the house lady had us checked out to make sure we were not thieves or the vice cops. I nearly passed out as I heard my nickname — "Pete!" — yelled out. Even though I had not seen Aunt Mary for more than twenty-three years, I recognized her right away. She was the after-hours joint's house lady.

I was her favorite nephew and she had spoiled me often with toys and clothes when I was around five or six. I had not seen many pictures of her over the years. Everybody there was now calling her Mary Smith. I enjoyed many wonderful weekends with her, eating free fish dinners and fried chicken sandwiches, breaking up card game fights and meeting many fine-looking Portuguese women before leaving for Camp Lejeune. I almost didn't want to go back to North Carolina.

It was kind of ironic. I was returning to Camp Lejeune with considerably more status than I had the last time I had been there, when I was charged with smoking an unlit cigarette. I sure wished I could have encountered Maj. Deshystler as a chief petty officer instead of as the naïve HM2 petty officer he grabbed by the arm back in 1963. As a chief, I certainly would have won that round. I ended up serving only one year of that scheduled two-year tour at Camp Lejeune. After a year, I was transferred to the Associate Degree Program at Tidewater Community College in Portsmouth, Virginia, where I earned an associate degree in education.

While at Tidewater Community College, I was highly recommended for promotion to senior chief petty officer and selection to Medical Service Corps officer. Despite the poor evaluations given to me by the Chief Nurse and the disciplinary actions taken against me by the command master chief in Viet

Nam. I continued to excel, but not advance. I received neither promotion to senior chief nor commission to Medical Service Corps Officer, even though I had one of the highest scores. I had the second highest score, behind Joe L Napier, among all of the hospital corpsmen in the entire Navy who were recommended for promotion to those coveted positions at that time.

After graduation, I submitted a request to attend Defense Race Relations Institute at Patrick Air Force Base in Florida. A personal written self-critique was required to accompany the application. I submitted my application, accompanied by the self-critique, outlining why I should be considered for the program, and was accepted.

****

To: Human Resources Management Program:
From: Fred T. Jackson HMC:
Subj: Application for selection to DRRI

I am applying for the Human Resources Management program because I will have a better chance to develop and utilize my potential. I will be bringing to the program a wealth of social experience in addition to my two years of college education. I have youth, the intelligence and the enthusiasm necessary to communicate effectively with, and relate to, today's young sailors. I also have the experience, maturity and military seniority essential to command the respect of my superiors and subordinates as well as my peers. Of my thirty-four years, eighteen were spent on farms in North Carolina, and the remaining sixteen in the Navy. As a sailor I have encountered many racial and cultural problems, all of which I have resolved or adjusted too successfully. I have demonstrated my potential by overcoming my limited, all-black rural education and performing all my assigned duties satisfactorily, by making rate thru E-7 in less than average time, by successfully passing the E-8 exam three consecutive times, by passing the officer selection battery twice, and by completing two years of college work in elementary education with better than "C" average. I've had a wide range of duty assignments and experiences ranging from ward orderly to field medical company aid man, to X-ray technician, to independent duty and division Chief Petty Officer supervising over a hundred and thirty man division in Viet Nam. My evaluations as CPO attest to my ability to communicate and counsel successfully across racial, ethnic and cultural lines. Finally, I am highly qualified for this program and feel I can vastly aid the Navy in achieving its human goals while significantly improving myself socially and advancing myself professionally.

HMC FRED T. JACKSON

LETTER TO THE EDITOR, NAVY TIMES
AUGUST 14, 1974:

While scanning the list of Z-grams published in your July 17 issue certain facts seemed to leap right out at me. In recent weeks I've listened time and time again to officers and senior petty officers express their obvious pleasure at the departure of the previous CNO and his cursed Z-grams. They placed great emphasis on how much better off the Navy is bound to be. No single Z-gram seems to be the source of their disillusionment.

Their denouncements are all inclusive. It is my personal feeling that a Navy with a CNO of Admiral Zumwalt's caliber is better than a Navy without a Zumwalt … I am privileged to have served in both.

Now to the facts: Your list contained 121 Z-grams. Three dealt with uniform regulations and grooming standards, two concerned themselves with equal opportunity, one with minority affairs and one with human resources management. From your letters over the years and from personal experience, these seven CNO DIRECTIVES repeatedly have been the major sources of most all the discontent. The wonder is that so few have prompted so much of it. Though there are only a few Z-grams dealing directly with human goals, all the rest are aimed in some way at improving the Navy or advancing human dignity. And those are matters with which all Navy people should be concerned.

What about those other 114 Z-grams? I suspect it would be rather difficult for any honest Navy personnel to scan your list and not come up with six or seven of those cursed Z-grams from which he or she has not derived a reasonable measure of personal benefit.

This brings me to the point that gripes me the most about many of my fellow bluejackets: They do not seem to have the courage to pick out the Z-grams they personally dislike and complain about them; instead, they generally denounce the whole package while enjoying the fruits of our previous Chief of Naval Operations' courageous efforts.

Finally, having never been CNO myself, I can only surmise that his job is just a wee bit bigger than that held by a MCPO, SCPO, CPO, etc. So it is expected he would make some mistakes in his decisions. But if the rest of us in the Navy made no more mistakes over the past four years than did Admiral Elmo Zumwalt, this great Navy of ours would be even greater.

HMC Fred T. Jackson

Getting selected to Defense Race Relations Institute started out to be a very interesting and rewarding assignment. After just a few weeks in the school six fellow students at the enlisted club robbed me during the final NBA basketball game between Boston and Milwaukee. The six men were covering my bets as I bet on Boston and we placed the bet money on a table between us. When Milwaukee lost to Boston, my classmates grabbed the bet money and ran out of the club. I ended up getting a court martial for attempting to recover my money at gunpoint.

I was removed from the program at Patrick Air Force Base and transferred to Naval Hospital Orlando, Florida, to await my trial, where I was convicted on two charges and served my sentence while awaiting my next assignment.

****

While serving at Orlando awaiting re-assignment, I accidentally encountered a gifted psychic named Alice Hull who made a profound and lasting impact on me. I call her a psychic for lack of a better word. She was amazing and I don't know if one should classify her as a medium, clairvoyant, psychic or something else entirely. She is undoubtedly the reason I stayed in the Navy as long as I did. I heard she has since passed away. Through the years I had extensive verbal and written communication with her and still retain her letters. She spoke very little about future events, though occasionally she would make a future projection. Mainly she spoke of things past, which thoroughly awakened me to long-forgotten or unknown situations, and circumstances, strongly reinforcing her abilities and completely freaking me out.

After the court martial and my sentence had been served, I decided to get my car worked on in preparation for any upcoming travel that might happen to accompany whatever orders I received. After doing some work on my car, the mechanic asked me to drive around and get about fifty miles or so on the engine and bring it back for his final tune-up. It was on a Saturday so I took a leisurely drive and ended up in a little village called Casedega and stopped for a beer. While having the beer, I recalled an article I had read about this little town in Saga magazine while on the plane coming home from Viet Nam in August of 1970. That article indicated the town was occupied by mostly psychics, spiritualist and mediums. I asked the bartender if this was that town and he said it was. When asked how I might get to talk with one, he said just wander around until you find one who is not busy and they will probably talk to you. So having some time to kill I left my car parked at his establishment and started walking around. There were two ladies sitting on the porch of the first house I saw. They told me they were waiting for a friend of who was being seen. I waited with them until their friend and Ms. Hull came out. I said I would like a visit with her. She looked at her watch and said she was expecting a delegation from the University of Florida, but had some time before they were due and invited me inside.

We sat down in her kitchen, which had black and white floor tile that I will never forget. I constantly counted those squares to make sure she could not read my mind. As she began, I was skeptical and expected her to ask me a bunch of questions. She didn't ask a single one. She began by saying let's see what the spirits give us today. Sometimes they give us things and sometimes they don't give us anything. She began after saying a short prayer.

The next thing she said knocked me completely over! She said, "Your mother, Laura, is standing behind me with her hand on my shoulder, and she wants you to stay in the Navy!"

How the hell did she know my mother's name and that I was in the Navy?

"Your mother wants me to give you the number 57," she said.

Well, that was my mother's name and she died at age 57. Needless to say, she had my attention.

She continued: "Your Aunt Mary gave you a nice sailor suit when you were five years old with a wooden whistle attached and you loved to wear it to church, walking from your three-room house barefooted."

I had completely forgotten about that sailor suit, which was given to me back in the summer of 1943, more than 30 years before. I had never thought of that house as having just three rooms. I always counted it as having five rooms. I realize now there were actually two areas I called rooms that were simply divided by curtains.

"You are thinking about getting out of the Navy now, but you should remain," she said.

How the hell did she know all this stuff about me!

She continued by saying, "Does the word Guam mean anything to you?"

"Not particularly," I said. "It's just an island in the Pacific and I've never been there."

"It will come to you," she said. "This Guam is not an island."

Continuing, she said, "Your sister Ellen who is in intensive care will be all right."

"I don't have a sister named Ellen," I replied.

She just simply said, "Oh, yes, you do."

I started to wonder what my father had been up to outside of his marriage. She went on and on for about a thirty minutes about so many things that seemed to ring a bell and so many that did not make any sense to me at all at the time. Finally, she hiked her skirt just above her knees and said, "This is where we stop."

I was totally caught off guard. How the hell did she know I

had been sneaking peeks at her pretty thighs while pretending to be counting the white and black kitchen tile?

"Here is my card if you feel the need to contact me in the future," she said.

She did not charge me anything but said if I cared to make a donation it would be appreciated. She really got my attention on a few things like my mother, the Navy, our three- room house and when I had the sailor suit so I gave her $20, thanked her and left.

On my drive back to the base I didn't reflect much on the things Ms. Hull talked about. Matter of fact, I forgot about most of them and reflected on the Saga magazine article I had read on the flight in August of 1970 and wondered if I could find a copy. I really wanted to read that article again.

Arriving back at the base after getting my car tuned up, I stopped by the Officer of the Day to check back in. He said, "Hey, Chief, I got a couple of things for you. Here is an urgent note to call your sister Johnsie in North Carolina. And, oh yeah, your orders just came in a few minutes ago to the USS Guam LPH 9. You are to report to the USS Guam (LPH 9) in Norfolk, Virginia, for duty. That's what Ms. Hull had been alluding to less than two hours earlier.

Needless to say the kinky Afro on my head stood straight up and I rushed to the pay phone in the hall to call my sister Johnsie. She said, "I have been trying to call you all day. Our sister Ella is in intensive care in Jersey City, New Jersey, and they don't give her much chance to survive."

I asked Johnsie what Ella's full name was. She said her name is not Ella. She got that nickname a long time ago. "Her real name is Ellen," she said. "We just always called her Ella."

I was completely floored. Ella recovered completely and lived for more than twenty more years. I was greatly impressed by Alice

Hull and stayed in touch with her from 1974 to the mid-1990s, when she wrote me a letter asking me not to try contacting her anymore. I tried writing to her anyway and the letter came back. I still have it and many others she wrote to me over the years.

****

I left Orlando for duty aboard the, USS Guam where I simultaneously relieved the medical officer and the senior enlisted hospital corpsman. Because of the many unjust disciplinary actions taken against me I was leaning heavily toward leaving the Navy. I am sure I would have retired had I not had the chance encounter with Alice Marie Hull on that Saturday afternoon in 1974 in Casadega, Florida. As it turned out, my assignment to the USS Guam was to be my last tour of duty aboard a Navy ship. It was by far the most challenging assignment of all my years in the Navy. It was even more challenging than the patient-handling assignment I'd had in the war zone in Viet Nam, which was really brutal.

****

"Anchors away, my boy. Anchors away! Farewell to college joys. We sail at break of day!"

When I reported for duty aboard the USS Guam I found the medical department in total disarray. The staff was made up of low-ranking medical personnel and there was a shortage of them. The division consisted of drug abusers, misfits, and members of the Church of Scientology and only a few of the required lab, operating room, pharmacy, X-ray and sanitation technicians. Additionally, I was given several collateral duty assignments, which included human resources and drug and alcohol abuse counselor.

The medical department was also poorly equipped and far below-required standards for medical supplies necessary for a deployment. I immediately set about the task of improving personnel, equipment and supplies and was able to adequately upgrade the medical division just in time for refresher training at Guantanamo Bay, Cuba. The final necessary medical equipment and supplies got loaded aboard as the lines were being cast off to get underway on that Saturday morning for our deployment to Mombasa, Kenya, in East Africa.

The USS Guam was an amphibious ship designed to carry assault Marines and their combat equipment, including helicopters, and Harrier jump-jet fighter planes. Harriers are fighter planes developed by the British with unique vertical takeoff and landing capabilities. My ship was designated to deliver a squadron of these fighters to Mombasa for further delivery to Jomo Kenyatta in Nairobi in time for their day of celebration of 13 years of independence from England. We got the planes to Mombasa in 1976 in time for Kenya's celebration.

Jomo Kenyatta threw one whale of a celebration and invited a delegation of sailors and Marines from our ship to attend in Nairobi. I was selected for the delegation but chose not to go because the delegation was required to fly from Mombasa to Nairobi. I had not flown since 1974, when I missed a flight from Florida by 30 minutes and it crashed in Charlotte, North Carolina. I had been on my way to my nephew Phil's funeral in Jersey City. He had lived with me before I left for Race Relations School and was to rejoin me after I completed it. Everyone on board was killed that day and I have not flown since.

The captain had arranged for us to fly from Mombasa to Nairobi so I allowed one of my subordinates, HM1 Phillips, to attend in my place. I really wanted to attend the historic celebration and meet

Mr. Kenyatta. I had long considered him a hero ever since I saw the movie "Something of Value" in which Sidney Poitier played a Mau Mau freedom fighter. Instead, I traveled from the port city of Mombasa by Land Rover vehicle to a game preserve outside Nairobi and enjoyed a full day of picture taking and sightseeing. The tour began immediately after reaching the preserve greeting station, where we were allowed to pet and take a few pictures with some tamed animals, including hippopotamus, giraffes and rhinoceros. After taking a few pictures there, we got a short briefing from the park rangers regarding the rules while visiting the preserve. Following the greeting session, we boarded three small Land Rovers and headed deep into the preserve to observe some true African wildlife. The first animal we encountered was a huge elephant alongside the trail. Our driver stopped so we could get some photographs and the elephant immediately charged toward us. Apparently the driver knew this routine would take place and got a real kick at us screaming for him to get the vehicle moving.

We proceeded toward the official eating area, eagerly taking photos of lions, baboons, cheetahs, skeletons, vultures and whatever else got in the way of our camera lenses. Upon arriving at the eating area, we were joined by two other groups of sightseers from Germany. Here we had a fairly good meal, which we shared with an awful lot of flies. The lunch consisted of burgers with all the usual fixings of lettuce, tomatoes, onions, pickles and cheese accompanied by strawberry drinks. This area was located atop a hill and provided an excellent view overlooking a river with thick jungle, tall grass, some brush and open patches on the far side. The opposite bank of the river was a good distance away, but with our zoom lenses we were able to get good photographs of a large herd of Cape Buffalo and some lions as we watched them go down to the river to drink. There were also viewing telescopes, with which we were able to get good looks at the vast surrounding area. I wondered how many TV episodes might have been filmed from this very spot. Quite a few, I expect, because the view was spectacular and the telescopes made it seem like we were right on top of the animals across the river.

After lunch we headed back to the ship docked at Mombasa. A few days later the ship returned to sea and headed back to the Mediterranean by way of the Suez Canal. The highlight of this voyage came when the ship sailed headed back to the Mediterranean by way of Alexandria, Egypt. While still in the Suez Canal, the ship got authorization from the chief of Naval Operations to stop for a few days in Alexandria. I will long remember that stop as one of the highlights of my navy travels.

Growing up in North Carolina, I had always been an avid reader; devouring every bit of reading material I could get my hands on. In high school, history and geography had been my favorite subjects. I often dreamed of visiting exotic faraway places I'd heard about or read about in books or magazines or seen in the movies.

Learning about Africa, Asia and the South Seas Islands stimulated my imagination and whetted my appetite for travel to faraway places. I often daydreamed of a time when I might be able to visit such places. My desire to travel was the No. 1 reason I joined the Navy; a close second was to avoid the draft. There were a few other things that also motivated me. Among them was seeing a guy named Gilbert Crockett in his sharp-fitting bell-bottom Navy blues with those thirteen buttons when he was home on leave and visited our school. He had been a tenth- grade dropout, but he sure looked mighty sharp in those Navy blues.

Boy! Did I ever travel after enlisting in the Navy! On this deployment to East Africa I found myself passing through the Suez Canal, crossing the equator, getting initiated as a shellback after having spent two exciting weeks in Africa, going on safari and photographing everything in sight. The thought of a visit to Egypt so soon after such a wonderful visit to Africa seemed too good to be true. I rushed to the ship's Special Services Office to see if any tours were being arranged. I was able to sign up for a tour to Cairo and the pyramids at Gaza. The tour lasted all day and was very fascinating. The mummies and historical artifacts in the museums really intrigued me. The mummified pharaohs appeared to be quite small. They did not seem to be of the same stature as the Egyptian tour guides walking around with us. They seemed more like children and I really wondered if such little people could have constructed and erected such gigantic objects as the pyramids and the great sphinx. The tour revealed much more than I had learned from reading books and the magazines collected from the abandoned house on Dewey Moore's farm.

I never imagined I would ever get to see Egypt in person. It would have been wonderful if we could have spent more time sightseeing along the Nile. Realizing this was probably my last sea tour; I had invested heavily in photographic equipment and sightseeing tours. It took many photographs and Super 8 movies.

****

In addition to the tours in Africa and Egypt, I participated in many sightseeing excursions when the ship returned to the Mediterranean. Sadly, my film was destroyed before it could be processed while we spent the mandatory period in Rota, Spain, before returning to the U.S. When Navy vessels operate in the Mediterranean, they are required to stop in Rota, Spain, to meet with the ships that are relieving them before returning to the U.S. Depending on the nature of the duty performed while in the

Mediterranean, the ships and equipment sometimes are required to be decontaminated. This includes descaling and thoroughly cleansing equipment of any possible infestation by foreign organism or pests.

During this turnover period, which usually lasts about ten days, it is customary for many sailors and Marines to send their photographic material to Brooklyn, New York, for fast processing so they can view and edit the items on the trip home from Rota. Sadly, X-raying during this debriefing period destroyed every bit of my unprocessed film. The only mementos I have left of this cruise are in my head and my ship's cruise book.

My entire military memoirs are accompanied by documentation in my possession, which helps immensely in jogging my memory these many years later. The only tangible evidence remaining of this trip in 1976 is my cruise book, which contains a brief history of events of the cruise and the administrative remarks in my service record attesting to the deployment.

I can describe many of the photos, movies and much of the activities and events I experienced on this voyage, but that would require so much more time and space to record from memory. I lost several Super 8 movies of bullfights in Spain, the leaning Tower of Pisa and many other historic sites, such as the Acropolis in Greece and the Roman Coliseum ruins. Everything was lost except the few souvenirs I purchased. I didn't buy many souvenirs this trip because I had been in the Mediterranean several times before. I had invested mainly in touring and picture taking. What a huge loss the destruction of my photographic materials turned out to be.

There were many more exciting events that took place once we were back in the Mediterranean. One incident involved the accident at Barcelona, Spain, where my ship had forty-nine killed and eighty-nine seriously injured. Twenty-four of those killed were sailors, while twenty-five were Marines. Twenty-four were white, twenty-four were black and one was Hispanic. I do have the records of the memorial tribute to those losses in my cruise book.

A quick trip to Google for news coverage of this incident involving the USS Guam LPH9 in January of 1977 should provide considerable information on the accident and news coverage for the several days we spent on recovery operations. There are a lot of interesting notes about this voyage contained in the USS Guam LPH 9 Association Newsletter, which can be found at www. USS Guam.org. Quite possibly you will discover some news footage of yours truly in operation in the midst of the recovery operation, while political rioting related to the movie "Victory at Entebbe" raged. On the way back to the pier after delivering some bodies to the airport, I was caught squarely in the middle of a major riot in front of a movie theater and barely escaped the barrage of bricks, bottles and bullets by ducking into the theater with the door guard.

I was directly involved in the recovery, treatment and shipping of all the dead home to their next of kin. I had gone into Barcelona earlier in the day and returned to the Guam before dark. We were anchored about two miles from port and I had allowed several of my medical corpsmen leave for having missed Thanksgiving, Christmas and the New Year holidays. They were still at the airport when the accident occurred, but I did not call them back from their leave in Norfolk.

Relaxing on my bunk while reading a novel, I was jolted upright by the intercom system. "This is not a drill! This is not a drill! All hands man your rescue and assistance details!" the voice said. I sprinted to my medical emergency station in my underwear and found my duty corpsman. "Chief, there was a collision in the harbor and the first boat with casualties is already alongside our ship!" he explained.

The captain had gone on a ski trip to Madrid, and I was several medical corpsmen short. The executive officer was left in charge. Before I could return to my quarters to get dressed, the executive officer was in sickbay asking for sedatives. He ordered me to report to the scene of the collision. I got dressed and organized, setting up shipboard triage for the casualties already starting to be brought aboard. After that was done, one of my men accompanied me on the rescue boat back to the harbor.

There was nothing my junior man could do at the scene to aid me so I sent him back to the ship because my medical crew was very short-handed back there. I had time to stop my seven medical corpsmen from flying out at the airport, but I opted not to cancel their trip. Had I known the extent of the casualties, I am sure I would have canceled the trip home for at least some of them.

The junior chief in my department, whom the executive officer kept aboard, was totally inept and I would not be back to the ship for a change of clothes for several days. The State Department men on the scene turned everything to do with casualties over to me. Never will I forget how they left me to my own devices in locating the injured as well as handling, identifying and shipping the dead servicemen's bodies back to the states. They promised to provide any support I needed. They gave me a business card with a phone number that no one at the embassy would answer and I never saw or heard from them again after that initial meeting on the pier.

I could not speak Spanish when visiting the hospitals and morgues searching for the injured and dead. I finally got a truck driver to take the first two bodies to the airport and paid him out of my pocket to act as an interpreter for me because he spoke very good English.

After the search was called off, I was able to return to the ship for a shower, hot meal and change of clothes. My junior chief back on the ship got himself designated the division chief while I was ashore, subsequently appearing as senior to me in our cruise book. I could have disciplined him for that move, but I chose to ignore it.

A visit to the U.S. Naval Archives and the official cruise book of the USS Guam for 1976 and 1977 will provide more details on this tragedy. There you can discover a great deal of interesting information about the entire non-stop trip from Norfolk, Virginia, to Africa and our return trip to America. That entire voyage was very enlightening with the visit to Africa and Egypt being the most memorable. I am still heartbroken each time I think about the loss of my pictures and movies taken on the most memorable of all my Navy travels.

****

My duty assignment after I left the USS Guam LPH9 turned out to be my last Navy assignment. It just happened to be in Charlotte, North Carolina, about 44 miles from where I grew up, in Anson and Union counties and my hometown of Wadesboro, North Carolina. From there, I retired after a 21-year naval career loaded with many eventful episodes.

While en route to and from Africa my medical department encountered and successfully managed many serious medical emergencies. Included in these were several appendectomies, a completely severed penis, forty-nine drowned and eighty-nine seriously injured sailors and Marines in the boating accident in Barcelona, Spain.

While on this deployment, I was again passed over for promotion even though I was highly recommended and awarded the Navy Achievement Medal for my outstanding performance during my assignment to the USS Guam and especially for the handling of the situation in Barcelona, Spain, after the boating tragedy. It seemed nothing I did could overcome the saga of the unlit cigarette.

****

The USS Guam's medical department was unfit for deployment when I reported aboard but highly commended when I left in 1977. Additionally, the Guam was dispatched to participate in several key operations in place of other LPHs because of the high state of readiness in which I left the medical department. That enabled the Guam to make many more assignments of importance before it was finally retired in 1998 and sunk as a target in 2001. Some of the deployments on which Guam was dispatched after I left, prior to being decommissioned, include the following:

- October 1983 – Casualty receiving ship during Grenada operation

- January 1986 – Space Shuttle Challenger recovery

- August 1990 – Operated in support of Desert Storm and Desert Shield

- January 1991 – Ordered to Mogadishu, Somalia, as part of Operation Eastern Exit

- June 1994 – Participated in 50th anniversary D-day celebration at South Hampton, England

- Spring of 1996 – Operated off the Coast of Liberia as flagship of Operation Assured Response

- October 1997-April 1998 – Conducted numerous exercises in the Persian Gulf in support of Operation Desert Thunder

My contributions to the Guam's medical department's operational readiness for these deployments cannot be over-emphasized and are noted in my evaluations. My notable contributions are clearly outlined in my Navy Achievement Award, the ship's commanding officer's evaluations, and the medical officer's critique of my performance.

After I left the Guam it was ordered to participate in quite a few important operations. USS Guam LPH9 was finally decommissioned on August 25, 1998. Three years later, on October 25, 2001, Guam was sunk as a practice target by the USS John Kennedy (CV-67) Battle Group.

I remain proud of the efforts I put in during my tour aboard and how they contributed to the USS Guam's final deployment capabilities. Upon the ship's return to the states, it was ordered to Newport News, Virginia, for an overhaul in 1977. While the ship was docked there, I took the opportunity to visit Washington, D.C., and the Navy Department. I wanted to see what assignments might be coming available for me upon completion of this, my last scheduled tour of sea duty. My rotation off the USS Guam was not very far off.

PART THREE

My Road Finally Takes Me
Back Home

****

CHAPTER 7

Home At Last

As I walked into the Medical Corpsman Detailers Office, wonder of wonders, the first person I encountered was my specifically assigned detailer who turned out to be a young sailor who had worked for me in the patient-handlers division in Viet Nam. I couldn't believe he was now equal to me in rank. I had outranked him by three pay grades when l left him aboard the USS Sanctuary in Viet Nam. Now here he was equal to me in rank and serving as my detailer.

I was very proud of his progress. He greeted me warmly by remembering how I'd been railroaded by Nurse Watson and the Master Chief of Deck Division. He then went on to say how grateful he was for the good evaluations I had given him in the patient-handlers division while I was his supervisor. He had been an outstanding corpsman when he worked for me in Viet Nam.

We chatted for a bit about the war zone days and patient handling. We had a good laugh remembering the time my brother, Bobby, a tough airborne soldier who was visiting me on the ship, tried to help us unload three helicopters filled with wounded during an emergency casualty receiving triage. Even though he was a combat soldier, Bobby had never seen so many bloodied troops at one time. He was totally knocked off his feet observing our triage efforts while trying to help unload a very heavy, very badly burned, very smelly soldier nicknamed "crispy critter."

I told my detailer that I had run into Ann O. Watson again at Newport, Rhode Island, and that she was still there when I left for Camp Lejeune. I wish I could remember his name but this was six years after Viet Nam and I'd had 132-plus corpsmen to supervise and evaluate back then. He said my good evaluations had gotten him assigned to patient administration when I left and that was where he worked until rotating back to the states.

He abruptly asked me, half-jokingly, "What can I do for you, Chief?"

I replied half-Jokingly, "Chief, you can get me transferred off the USS Guam and assigned to my hometown."

He shuffled his papers briefly, looked up at me and said, "Okay, I've got just the thing for you. I actually have an opening coming up in Charlotte, North Carolina, at the Navy and Marine Corps Reserved Center about 44 miles from your hometown and can rotate you off the ship in about six months. Would you be interested?"

****

"Country roads, take me home to the place where I belong."
–John Denver

I rotated off the USS Guam to report to Charlotte, North Carolina, in 1977.

****

"Hey, it's good to be back home again. Sometimes this old farm seems like a long-lost friend."
–John Denver

I served two years and three months of the assigned four-year tour in Charlotte. While there, I got turned down two more times for higher enlisted promotions and officer selection but was offered a transfer to Hawaii. By now I was becoming disgruntled and more and more disillusioned at being forced to remain in one grade for so long.

I was clearly qualified for promotion and constantly receiving outstanding evaluations from all the senior officers under whom I served in spite of the negatives entered in my record during the first ten years of my naval service. (See Chapter 10 on Cheers)

I decided to apply for retirement. It was granted and on January 29, 1979, this memorable No. 1 trip came to a close with all of the many fears, jeers, cheers and tears I encountered indelibly stamped into my memory for evermore. I had a total of 21 years of naval service, with the last eleven years spent mired in a single pay grade. This was highly unusual, and I was very disappointed to witness so many of my peers of lesser qualifications, as well as so many junior sailors I trained, getting promoted ahead of me.

****

****

CHAPTER 8

Tears

"A reform is a correction of abuses; a revolution is a transfer of power."

- Edward George

Thanks again, unlit cigarette, for denying me my third good conduct award and preventing me from qualifying for and receiving the coveted gold stripes. Gold stripes carried so much more weight than red ones when being reviewed for promotion to senior enlisted rank or officer selection.

I wish I could recall my last detailer's name. He is one of only two people that stand out as having looked out for me and treated me fairly when it counted during my career in the Navy. The other was the administrative officer at Naval Hospital Orlando, Florida, who supervised my court martial before and after I was sentenced. In order to get information on them requires their permission in accordance with the privacy act. The administrative officer was an enlisted man who rose through the ranks to become an officer. He kept me from being confined and doing hard time. He helped me regain my chief petty officer ranking and maintain my dignity. He even gave me an outstanding transfer evaluation once my orders to the USS Guam arrived. Even though I exercised poor judgment at Patrick Air Force Base, he recognized the provocation and the injustice of how I was convicted on the testimony of the very service members who had robbed me.

I began the steps of retracing this journey by requesting my complete military record 33 years after retiring from the Navy and 54 years after enlisting. I thought it would be interesting to review the administrative remarks pages first. What an eye opener that turned out to be! I was unashamedly moved to tears as I began my review.

CHAPTER 9

Jeers

Jeer as defined by Webster's dictionary: "To deride, to mock, to treat scoffingly, and to inflict with railing remarks."
"Many commit the same crime with a different result. One bears a cross for his crime; another receives a crown."

- Juvenal

"I'd like to hold my head up and be proud of who I am, but they won't let my story go untold. ... I've paid the debt I owe them but they're still not satisfied; now I'm a branded man out in the cold."

- Merle Haggard

While reviewing and itemizing the various disciplinary actions taken against me I was able to understand the quote by Juvenal. While given the eleven crosses to bear for crimes I did not commit, I watched others commit far more grievous infractions and receive no punishment whatsoever. The following disciplinary actions taken against me had a cumulative and very negative impact on how far I was able to advance, no matter how well I performed.

- UCMJ ART 92: Failure to attend training class – Pled not guilty. My ID showed I was already promoted beyond the requirement to attend. Disposition: Convicted, awarded punishment.

- UCMJ ART 86: Unauthorized absence from place of duty – Pled not guilty. My name was added to the duty roster after I had already been dismissed for the day. Disposition: Convicted, awarded punishment.

- UCMJ ART 86: Failed to report for chow relief – pled not guilty. I was assigned as chow relief after I had been dismissed for the day and left the base. Disposition: Convicted, awarded punishment.

- UCMJ ART 92: Failure to take charge of cleaning assignment – Pled not guilty. Completed task after dinner with five hours to spare. Disposition: Convicted, awarded punishment.

- UCMJ ART 117: Wrongful use of provoking words – Pled not guilty. Informed the barracks master at arms I was going to chow and would clean the area when I returned. I outranked him by one pay grade. Disposition: Convicted, awarded punishment.

- UCMJ ART 86: Failure to obey a lawful order – Pled not guilty. Ordered to leave the Field House and remove a cigarette from my mouth, I did both as ordered. Disposition: Convicted, awarded punishment.

- UCMJ ART 134: Creating a disturbance on a boat– Pled not guilty. Requested the coxswain not to cast off before the three approaching chiefs reached it. I outranked him by four pay grades. Disposition: Convicted, awarded punishment.

- UCMJ ART 92: Failure to obey a lawful order – Pled not guilty. Directed to go to my quarters, I did so after stopping for a sandwich and coffee in the CPO Mess. Disposition: Convicted, awarded punishment.

- UCMJ ART 128: Assault – Pled not guilty. I was punched in the mouth by one of my subordinates whom I took measures to restrain. I outranked him by three pay grades. Disposition: Convicted, awarded punishment.

- UCMJ ART 128: Assault – Pled not guilty. I was robbed by several classmates at Race Relations

School and recovered my property at gunpoint.
Disposition: Convicted and awarded punishment

- UCMJ ART 134: Careless discharge of firearm –
  Pled guilty. Discharged the weapon into the floor
  after recovering my property from the six men who
  had robbed me at the club. Disposition: Convicted,
  awarded punishment.

CHAPTER 10

Cheers

Citations – Certificates – Medals – Award

"I have learned that success is to be measured not so much by the position one has reached in life as by the obstacles which one has overcome while trying to succeed."

- Booker T. Washington

Following are some of what I refer to as cheers. I call them that because I am cheered up and get a good feeling when reviewing what I accomplished in spite of all the jeers faced. The lowliest certificates are equal, in my mind, to the highest evaluations or awards filled with platitude and accolades. The highest of them all being the Navy Achievement Medal awarded after my final cruise aboard the USS Guam LPH-9.

There is no doubt that award could have been higher. I am certain at least a Navy Commendation Medal would have been appropriate. I understand why the tragic accident in Barcelona was not dealt with or mentioned in my commendation. That would have played up negatives about the cruise and reflected poorly on the ship even though the ship actually had nothing to do with causing the accident other than allowing the Liberty Launch to be overloaded.

Especially inspiring to me are the many great evaluations earned and received during the last ten years of my service. It is difficult accepting the fact that an unlit cigarette could have been so effective in neutralizing those outstanding performances, evaluations, accomplishments and awards. But it sure did. Despite the best cheers of my career I would never receive any more promotions. In the end, I was also denied a well-earned retirement ceremony.

I can, however, see why the Navy often sought my expertise as documented in my evaluations, and why they utilized me in so many difficult problem-riddled situations. And I managed each of them successfully, I might add. At the same time, I am hard pressed to understand why the appropriate rewards did not come along with such good evaluations and highly acclaimed achievements. Perhaps reasons existed why they did not lead to more promotions or commissions. But when you consider the entirety of my records there is no justification for it. The main factor was me not being able to get a third conduct award and earn gold stripes due to the unlit cigarette incident.

I sure would like to get down into the weeds though and figure out what I did or didn't do that caused me to get stalled in a pay grade for ten years while receiving such impressive and outstanding evaluations. Such good performance evaluations as mine are not generally conducive to promotions failure. Good performance, outstanding evaluations and persistence had always been adequate to overcome whatever negatives my records contained during the first ten years of my career.

The unlit cigarette has to be the missing piece of the puzzle that combined to stymie my promotions. Its impact on my life cannot be overemphasized. During my last ten and a half years in the Navy, I never got another promotion even though those were the years I received my most outstanding evaluations and was given the most responsibility.

With such good performance and outstanding evaluations on record, I am bewildered and completely mystified why my promotions came to a complete halt and the negatives could no longer be overcome from 1969 to 1979. As I review my many cheers, in the vernacular of an old card player, all I can do now is "read 'em and weep." And sometimes I do just that!

Reading over this list of cheers, it is hard to see how promotions for me dried up so completely. Items #1 to #32 below seem to reinforce my eligibility for continued upward progress.

- 08/28/1958 – Certificate of Completion – Hospital Corps School

- 12/16/1958 Certificate of Promotion to – HN (E3)

- 12/16/1959 Certificate of Promotion to – HM3 (E4)

- 10/30/1960 Certificate of Completion – Field Medical Service School

- 11/16/1962 Certificate of Promotion – to HM2 (E5)

- 03/16/1964 Certificate of Completion – Medical X-ray School

- 04/16/1966 Certificate of Promotion to HM1 (E6)

- 05/05/1966 Awarded National Defense Service Medal

- 06/09/1967 Certificate of Completion – Drug/ Alcohol Program Advisor

- 08/16/1967 Certificate of Completion – Advanced Hospital Corps School

- 04/10/1968 Awarded Good Conduct Medal #1

- 07/16/1969 Certificate of Promotion to HMC (E7)

- 12/14/1969 Awarded Meritorious Unit Citation with Cross of Gallantry – USS Clarke County LST 601

- 02/01/1970 Awarded Meritorious Unit Commendation USS Sanctuary AH 17

- 09/08/1970 Awarded Viet Nam Campaign Medal

- 09/08/1970 Awarded Viet Nam Service Medal

- 01/01/1971 Certificate of Completion – American Registry of Radiologic Technologists

- 05/19/1974 Awarded Associate in Science Degree – Tidewater Community College, Portsmouth, Virginia

- 05/19/1974 Awarded Good Conduct Medal #2

- 04/03/1975 Certificate of completion – Career Counselor School

- 08/12/1975 Certificate of Completion – Management of Sexually Transmitted Diseases

- 09/19/1975 Certificate of Completion – Career Information and Counseling

- 05/16/1976 Awarded Navy Achievement Medal

- (24). 01/31/1974 Commanding officer, naval hospital, Camp Lejeune, N.C. Endorsement and transfer Evaluation for Fred T. Jackson: Chief Jackson has discharged his responsibilities in a highly creditable manner. He has been especially effective in counseling the corpsmen assigned to his direction and in maintaining good patient contact in this busy department. He has designed and introduced an effective accounting system for the department's workload. Career motivated, he has pursued successful off-duty study and correspondence courses to improve and enhance his advancement opportunities. He has demonstrated no serious deficiencies in this department and it is anticipated he will perform well in the future. Promotion is recommended. W.A. Harrison, Capt/MC USN

- (25). 12/19/1974 Commanding Officer, Naval Hospital, Orlando, Florida: Transfer Evaluation of HMC Fred T. Jackson: Rate has far surpassed expectations in all categories of evaluation, especially Performance and Conduct. He was transferred to this command with disciplinary action pending from his previous command, DOD Race Relations School. He was found guilty at a special court martial of violation of articles 128 and 134, sentenced to 30 days restriction, forfeiture of $150 x 2 months and suspended reduction to E-6. His Performance at this activity has been far above average. Even with his obvious concern for his Navy future, he has performed assigned tasks in a credible manner. His personal appearance and demeanor has been excellent. He has displayed enthusiasm and initiative in each assigned activity. HMCM O'BRIEN, ADMINISTRATIVE OFFICER BY DIRECTION

- (26). 5/15/1975, COMMANDING OFFICER, USS GUAM LPH-9: FIRST SHIPBOARD EVALUATION OF HMC FRED T. JACKSON; Chief Petty officer Jackson has performed all duties assigned in an exemplary manner, and in addition has volunteered his services as RAFT Counselor. Since detachment of the ship's medical officer he has acted as head of the Medical Department. A strong and forthright person, Chief Jackson has kept the Guam's Medical Department together with no loss in service to crewmembers since the Medical Officer's departure. He has recommended several actions in dealing with drug and alcohol abuse problems, all of which appear successful. As RAFT Counselor he has measurably increased the effectiveness of Guam's phase ll Human Resources Program. An intelligent, articulate man, his use and knowledge of the English language, both written and oral, is exceptional. He is highly recommended for promotion to Senior Chief Petty Officer. T. A. STANLEY/CAPT USN

- (27). 5/16/1976 COMMANDING OFFICER, USS GUAM LPH-9 SECOND SHIPBOARD EVALUATION OF HMC FRED T. JACKSON: HMC Jackson has earned the HM8452 NEC for Independent Duty Hospital Corpsman. During this period he has steadily improved his excellent trend of performance to outstanding. He has administered his diverse programs exactly in accordance with published guidance. Working with minimally motivated and capable assistants he has consistently improved every aspect of the hospital service aboard. Doctors temporarily assigned during at sea periods, consistently praise his strong willed dedicated performance. Chief Jackson is a gifted instructor, appearing on onside

T. V. training with each division; he has trained the entire crew in the required areas of first aid and nuclear weapons radiation hazards. As the primary Rights and Responsibility workshop facilitator he has significantly and directly contributed to the command's affirmative action plan and the equal opportunity programs, as well as promoted increased responsibility. In comparison with his contemporaries, Chief Petty Officer Jackson stands 1 of the 3 so rated in item 26. T. A. STNALEY, CAPT/ USN

- (28). 6/16//1976 COMMANDING OFFICER, USS GUAM-LPH9 PRE-DEPLOYMENT EVALUATIONJ OF HMC FRED T. JACKSON: HMC Jackson is the Division Officer for the medical department. Additionally, he has collateral duties as division training petty officer, division career counselor, assistant ship career counselor, and member of the Human Resources counsel and facilitator for military rights and responsibility workshops. HMC Jackson has done an outstanding job welding together an effective medical department. This is especially true since he has had to eliminate some non-performers and then develop new men fresh out of the respective technician schools. He is a very capable administrator and knows well how to motivate his men. While I was the assigned Medical Officer, HMC Jackson's achievement in upgrading the medical department has been outstanding.

Previous disappointments have been present with
no medical officer aboard. He is an outstanding
example and his untiring work during this portion
of the cruise resulted in marked improvements
in the medical department. In view of my
observations of this HMC, I strongly recommend
him for advancement to E-8. I would be proud
to serve with him again anywhere. WALTER
BEASLEY, CAPT/MC USN

- (29). 11/16/1976 COMMANDING OFFICER
  USS GUAM LPH-9 TRANSFER EVALUATION
  OF HMC FRED T. JACKSON: HMC Jackson's
  responsibilities have been awesome during this
  period. He has performed in a superb manner
  in every respect at a level well above that held.
  His dynamic leadership and aggressive, effective
  supervision has assured the medical department
  met and passed all inspections in an excellent
  manner. His department and ship-wide training
  insured that several serious medical emergencies
  were met successfully. His department had
  the lowest disciplinary problems, the highest
  promotion rate (100% E-4 thru E-7), and the
  highest retention rate. His entire department
  was the first to attain 100% PQS Qualification
  in damage control. By following his example of
  superior performance, 2 of his corpsmen won
  sailor of the quarter. With embarked units he
  supervised 34 regularly assigned and embarked
  medical corpsmen. His well-planned and persistent

follow-up resulted in massive upgrading of the material condition of the medical spaces and equipment reliability. Managing a vast Authorized Medical Allowance list he has, through initiative, maintained absolute control over medical supply inventory and kept it near 100%. During the deployment to east Africa and the Mediterranean the medical department's ability to support various medical officers and surgical teams were repeatedly demonstrated. Additionally, Chief Jackson has been an inspirational leader, establishing and conducting extremely effective human goals programs, facilitating workshops in military rights and responsibilities as well as cultural expression in the Navy. Chief Jackson's positive accomplishments readily attest to a high degree of professional capability to lead, manage, counsel, direct and motivate personnel individually and as a team. He is strongly recommended for promotion to Senior Chief. T. A. STANLEY, CAPT/USN

- (30). 11/16/1977 COMMANDING OFFICER, NAVY AND MARINE CORPS RESERVE CENTGER, CHARLOTTE, NORTH CAROLINA ANNUAL EVALUATION OF HMC FRED T. JACKSON: CHIEF JACKSON, IN ONE YEAR'S TIME, HAS DEVELOPED THIS COMMAND'S MEDICAL DEPARTMENT FROM THE LOWEST DEPTHS OF INADEQUACY TO A PROFESSIONAL, RESPONSIVE AND WELL

OPERATING MEDICAL UNIT. He has brought this navy command a renewed respect for the navy medical branch, and through his technical and leadership expertise, the personnel assigned, active and reserve, are again confident that they can be provided all available medical resources when required. Chief Jackson promotes the navy's equal opportunity programs and his actions are an outstanding example of the "Navy takes care of its own" motto. RICHARD A. THRELKELD, CDR/USNR

- (31). 6/16/1978 COMMANDING OFFICER, NAVY AND MARINE CORPS RESERVE CENTER, CHARLOTTE, NORTH CAROLINA REGULAR ANNUAL EVALUATION OF HMC FRED T. JACKSON. Chief Petty Officer Jackson has proven to be a highly valuable member of this command. He brought an exceptional level of expertise to the medical department that had been sorely lacking. He has spent many hours in the challenging task of upgrading and certifying the medical records of over 400 drilling reservists, in addition to the active duty staff. In many cases, it was necessary for him to reconstruct entire records for those personnel who had lost or missing records. In view of the fact there is no active medical staff or physicians assigned, other than Chief Jackson, this was a monumental task that was accomplished in an enviable manner. His reliability, resourcefulness and initiative were highly commendable. He needs no direction whatever in carrying out his assignments and often anticipates requirements well in advance so that the medical department operates on a smooth efficient plane. Chief Jackson has been especially helpful with the command's equal opportunity program and has effectively provided beneficial counseling and advice to minority members of the command. Of all the challenges Chief Jackson has faced, the greatest has been to serve as acting officer in charge of the reserve medical unit assigned to the center. Lacking direction and a strong leader, the unit was floundering in misdirected training programs and accomplishing very little. Under Jackson, a meaningful program has been developed and the unit now has a sense of purpose and direction that has greatly improved morale. Through his efforts with this unit, he has amply demonstrated those leadership talents so necessary in a chief petty officer; Chief Jackson is a definite asset to this command and would be an asset wherever assigned. He has the potential for increased responsibility and is highly recommended for advancement to Senior Chief Petty Officer. W. H. THRELKELD, CDR/USNR

- (32). 6/16/1977 COMMANDER, NAVAL SURFACE FORCE, UNITED STATES ATLANTIC FLEET NAVY ACHIEVEMENT MEDAL AWARDED TO HMC FRED T. JACKSON: For professional achievement in the superior performance of his duties while serving as "H" Division Officer aboard the USS GUAM LSPH-9 from 28 March 1975 to 30 December 1976. Chief Petty Officer Jackson consistently performed in an exemplary and highly professional manner. Displaying exceptional professionalism and resourcefulness, he relieved the regularly assigned Medical Officer and Chief hospital Corpsman simultaneously and effectively carried out the duties of both division officer and senior medical representative. Through hard work and dedication, he corrected numerous discrepancies which contributed greatly toward the improved operational readiness of the USS Guam's Medical department during his ship's special deployment to East Africa, Chief Petty Officer Jackson's technical skill and guidance enabled the Medical Department to effectively prepare and support surgical teams capable of operating under difficult conditions at sea. His diligent efforts and competent leadership inspired all who observed him and contributed significantly to the accomplishment of his ship's special mission. Chief Petty Officer Jackson's exceptional ability, initiative and total devotion to duty reflected great credit upon himself and the United States Naval Service. W. L. READ, ADMIRAL/USN

PART FOUR

The Good, The Bad, The Ugly & The Memories

****

CHAPTER 11

Truth or Consequences

NBA Exihibition Game, Field House, Marine Corps Base Camp Lejeune, North Carolina, 1963

I said, "Yo, Jenkins! It's less than a minute to halftime. Let's step outside and have a smoke."

"Hey you!" someone yelled. "Get that cigarette out of your mouth!"

"Man," I said, "the cigarette is not lit. I'm going to light it after I get outside."

He repeated, "I said, 'Take it out of your mouth now!' "

And I said, "Man, who the hell are you? I don't have to take the cigarette out if it ain't lit. "Come on, Jenkins, let's step outside."

"All right, you are under arrest for disobeying an MP's direction."

"Man, I said, 'who the hell are you?' Let go of my arm."

He said, "I am Major Dishystler, officer in charge of the MP Company, and you are under arrest. Write him up, private."

At the time I could not imagine the long-term repercussions

this incident would have for me, including the loss of gold stripes, loss of future promotions, loss of creditability, loss of good conduct award, rejection by senior enlisted and officer selection boards, and short- and long-term loss of income.

In retrospect, as I moved up in rank, I became fully aware that had I been a white petty officer this incident, and probably many of the other charges, would never have been brought against me. If they had, I would have been given a fairer hearing and evaluation of the facts than the Marine major and my commanding officer gave me. I have never understood why my commanding officer at the hospital sided completely with the Marine major, totally ignoring the facts of the case and my eyewitness, PFC Jenkins. The major who ordered me arrested had spent the entire game standing in civilian clothes near where most of us black troops were seated. When he identified himself, we were already out of the gym. The cigarette was out of my mouth and had never been lit. The punishment was most definitely excessive and unjustified. There were many other witnesses besides PFC Jenkins, none of whom my commanding officer bothered to seek out.

CHAPTER 12

Let the Truth Be Known

'The victor will never be asked if he told the truth."

-Adolph Hitler

In reviewing the many "cheers" I've listed here, it would appear I overcame the many "jeers" entered in my record. But I didn't, and this concludes the overview of looking back at my military career and the life, which propelled me to it. The major effects an unlit cigarette had on my career cannot be overstated.

This narrative has numerous episodes contributing to these many stories. There are some funny ones, happy ones, curious ones, short ones, long ones. There are descriptions of failure, of pain and success. There are stories I wish never to forget and stories I wish I could, but probably never will. They are all here, and based on my military records, and my childhood recollections. Thanks to you, oh mighty unlit cigarette, for derailing a military career filled with so much promise and potential! Every sailor has at least one story to tell. These are some of mine, and I'm sticking to them. There is no intent to prove innocence here. No one is likely to think I could be innocent of so many charges, but I was. Hopefully, this memoir will interest readers enough to investigate for themselves. In every case but one, there is much to indicate I certainly was not guilty as charged.

There are facts and many documents in support of my narrative. It's up to whomever has interest to research and review the documents and then arrive at their own conclusions. Whether or not my stories are accepted as fact, imagine what a struggle it must have been for me to try to overcome the long-term negative consequences. Especially, if I was, in fact, wrongfully charged and disciplined unjustly.

Finally, this brings me back to my father, who passed away and was laid to rest on March 26, 1983. As I listened to the choir singing his favorite hymn, "A Little More Faith in Jesus," I was able to view my return to civilian life with a fuller appreciation of all he had done on his family's behalf. I deeply regretted him not getting to benefit very much by justice gaining the upper hand over injustice. Today, I am still saddened that I lived these years described in this memoir without the full benefit of justice's triumphs. Certainly, if I were to begin my journey today, rather than back in 1939, my trip to the top would obviously be significantly easier.

Even so, I am grateful that my children and grandchildren will have a much less rocky path to navigate as they pursue their goals. Opportunities denied to my father and me are becoming increasingly available to them as justice continues to subdue injustice and its effects. Though justice does not seem to have totally overcome hate and prejudice, I believe it is certainly gaining and sustaining the upper hand over injustice every day.

Just prior to retiring from the Navy in January 1979, I discovered my father was suffering from Alzheimer's disease and no longer knew who I was. Looking back, I remember him fondly as I've attempted to compare and contrast our journeys. I finally realized Daddy never moved just because he had to. He always moved because he thought it was best for the well-being of his family. What a great man he was! All my frequent moves were mostly forced on me solely for the well-being of the Navy, with little regard for whether I wanted to move or not.

I came to realize it was my father's example of passive dignity that helped me persevere and refrain from rebelling and becoming totally militant instead of trying to get the most I could from the Navy. I wish I had been more forceful in defense of myself, but, like my father, I always thought performing and achieving at a high level would overcome things.

Culturally, I was conditioned by Jim Crow and allowed myself to be completely bowled over by the injustice of covert racism practiced during my entire career. By the time Malcolm X, Stokely Carmichael, Rosa Parks, Martin Luther King and other black leaders came to prominence and ignited a militant spark in me, my military record had been effectively predestined for unfair treatment and James H. Crow, Junior Esq., had replaced Jim Crow.

Along with many other socially deprived and economically disadvantaged young sailors, I was set up for failure due to long-standing institutional racism and bigotry. Discrimination was blatantly practiced throughout my entire Navy career. But the example of my father's tenacity and perseverance fueled my desire to excel and made it possible for me to endure the many injustices I would repeatedly encounter. My father's example also would not allow me to consider failure as an option.

My brother Eddie will always be remembered as the person who taught me to never call a job finished before you had tried in every way to get it done, even if it was not done right, always finish it he said, and let someone else decide if it was done right. But always get it done completely before giving up.

Ms. Smith-Patton motivated me to try to read and speak well, and encouraged me to read the classics such as "The Robe" and the "Rhyme of the Ancient Mariner", in addition to the comic books I kept squirreled away in my desk.

Ben Morgan forced me to look at life through a philosophical prism enabling me to evaluate the various episodes of my entire life in such a manner that the seemingly unbearable ones could be managed, overcome and made acceptable at a satisfactory level.

It was impossible for the civil rights movement alone to help me totally salvage my military record in light of all the gross injustices I faced while I serviced. I was amazed that I rose as fast and as far as I did in the Navy, until I was able to review my was amazed that I rose as fast and as far as I did in the Navy, until I was able to review my

entire military record of performance, and reflect back on the contributions of my brother Eddie, Ben Morgan, Ms Smith-Patton and my father. I am sure there were others but these four are clearly standouts in my mind.

CHAPTER 13

To Quit or Not to Quit

*"The devil is strong in me. ... Rebellion is in the blood, somehow
or other. Now I can't go on without a fight."*

- Henry Brooks Adams

My retirement ceremony was in line with all that had gone
before me during my service in the Navy. I was not given a
retirement ceremony and, therefore, denied the traditional pomp
and circumstance other chief petty officers got of symbolically
being piped over the side of a ship as a final tribute. I had
participated in many retirement ceremonies over the course of
my military career and was very much looking forward to mine.
Being denied that final tribute from my shipmates was truly a
huge disappointment. If you review my evaluations for the last
ten years of my service, you will be hard pressed to say I did not
deserve a CPO retirement ceremony.

The reason given by the Navy was logistics. I was stationed
over a hundred miles from the nearest Navy base when my
retirement date approached. I was told that transferring to

Charleston to await my retirement was not practical. I was needed at my current duty station in Charlotte, North Carolina, until my relief arrived. This was total unadulterated bullshit! After all I'd been through; this seemed like a final slap in the face, punctuated by a kick in the groin. Reflecting on the outstanding evaluations of my final ten years in the Navy, it was, and still is, extremely difficult to understand how I could have been denied promotion in the 1978 November or December cycle, or a retirement ceremony in January 1979.

"The roots of my raising run deep. I've come back for the strength that I need, and hope comes no matter how far down I sink. The roots of my raising run deep. A Christian Mom who had the strength for life the way she did then to pull that apron off and do the Charleston for us kids. Dad a quiet man whose gentle voice was seldom heard and who could borrow money at the bank simply on his word."

-Merle Haggard

EPILOGUE

After being released from the Navy at Charleston, South Carolina, on January 29, 1979, the drive back to Charlotte was bittersweet. The entire trip was spent reviewing my Navy career as if it was a video recording on permanent rewind in my brain. Sometimes I found myself smiling and chuckling, sometimes scowling, and sometimes with my eyes welling up with tears.

I did not yet feel a need to seek revenge, retribution, or to strike out in any way. The thought of fighting against some sort of injustice would not come for thirty-three more years when I received a copy of my military records. By then, justice was out of the question. I had always felt I got vengeance enough each time I overcame an injustice by securing another promotion. I am inclined to feel that was the right way. That's how Daddy would have done it. That seemed to be adequate retribution until the promotions dried up as my qualifications grew stronger and stronger, continually reflecting my outstanding performance of duty.

After processing out at Charleston, South Carolina, it seemed like no time at all passed before I found myself back home in civilian clothes, but still very much cloaked in the military state of mind developed over the past twenty-one years. I sat down in my den, looked around at all the military memorabilia on the walls, tables and desks, and then set about the task of debriefing myself on more than two decades in the military and the previous eighteen years spent in rural North Carolina. It was amazing to realize I was just 39 years old after everything I had experienced.

I wondered what had become of the three young white playmates I'd had between the ages of three and seven. The nearest black kids I had to interact with during this time were at least four miles away, and I only saw them regularly after I started school or at church on Sundays.

Carolyn Thomas was my playmate at age three. Charles Coleman and Henry Jay were my playmates from age four to six. I wondered if I had made some kind of lasting impression on them like they had on me, how life had treated them, and what they might be doing after all these years. Pondering and reminiscing I couldn't help but consider retiring to the plot of land my sister Johnsie had given me when I saved her ten acres near Marshville, North Carolina, from being auctioned for delinquent taxes back in 1964 while I was stationed at Camp Lejeune as an HM2 — and smoking unlit cigarettes.

Marshville is a little town along Highway 74 in Union County, North Carolina, where I had spent several years growing up. It is also the place where I boarded the bus as I left to enlist in the service. Johnsie, my oldest sister, had worked as a cook at The Wagon Wheel, Little Tom's, and the Palomino Grill for more than 40 years in Marshville. The Palomino was owned by an actor who had played The Cisco Kid's sidekick in the movies and on TV. As a matter of fact, his step-son, David "Chico" Barrett, had been stationed with me for a while in Charlotte and we had become good friends. David loaned me the start-up capital to start a small janitorial company after I retired from the Navy. I eventually grew that business, Complete Cleaning Company, to 53+ full and part-time employees.

My sister Johnsie was well known and well respected around Marshville. She once had her picture taken with Randy Travis, the country singer, while she was working at Little Tom's. A tribute was placed near East Union School commemorating Marshville as the hometown of Mr. Travis.

As I was wiping away tears, I wondered could there still be more to come. There was, but I would be 72 years old before I thought much more about it. That happened when I decided to send for my military records. I received them in July of 2012. My records put in perspective for me, for the first time, the big picture of the twenty-one years I spent in the Navy and the effect my 18 years of rural cultural conditioning played in my development.

It became clear to me that life is not easily scripted. I was able then to try reconciling my military career with the rest of my existence, before and after the Navy. Hopefully, I've began to accomplish that with this memoir. With that final thought, I will be quite pleased if anyone, other than me, finds this interesting enough to read. If they do, perhaps they will understand this is not a family history, but an honest effort to show how family and the military contributed to whom and what I have become. I had one daughter born while I was in Iceland and another while I was serving aboard the USS Notable MSO 460 in Charleston and got to spend very little time with them until I retired and my wife deserted them. They came to live with me for some time and we grew very close.

After the divorce my ex-wife let the children visit with me quite often. They came to visit me at Newport, Rhode Island, Norfolk, Virginia and Camp Lejeune. We had lots of good times going to movies, beaches, picnics and rodeos. One visit I remember with particular fondness was when I was able to take them out on my ship with me during a family day outing. Capt Stanley let Tomiko, who was nine years old at the time, sit in his chair on the bridge and pretend she was steering the ship as we slowly steamed in the Chesapeake Bay.

Tomiko was extremely proud of herself thinking she was driving the ship. Kym, who was seven at the time, was not very much impressed. She was thoroughly mesmerized by the magic show we attended later that night at the Marine staff club on Hampton Boulevard.

They eventually went to live with their mother until they left for college and the army reserve. Tomiko went into the army reserve for a brief time and Kym, the younger one, got degrees in law and journalism, married a career Marine and eventually went to work for the governor's office in Atlanta.

Tomiko eventually started a couple of businesses in the Atlanta area. I enjoyed them during the short period of time they lived with me and they seemed to enjoy me too. I will always remember when they went back with their mother and how I found over a hundred dollars' worth of overdue library books under their beds. We all obviously shared a love of reading

As I close this, I am reminded of a quote from Dante's Inferno: "Wretched souls are those who have lived without praise or blame." Without a doubt, I received a good measure of both praise and blame, duly taking note of their impact on my life.

Final note from author:

Looking back over my rural upbringing, I find it difficult to find anything I would do different if I could, except maybe avoid that doggone copperhead snake. The unlit cigarette charges would not be accepted without a courts martial to hear the evidence.

I have many good memories of my association with that little five room wooden high school at Polkton where I graduated in 1957. It never ceases to amaze me when I reflect upon the many accomplished persons who attended that school after me and before it was closed and became a consolidated county school in the early 1960s. One might be surprised at the many outstanding people developed from among us barefoot, so-called, deprived and economically disadvantaged colored folks who attended there. From that school came many accomplished peoples, including PHDs, lawyers, scientists, and high ranking military officers, including a three star general. I will forever be grateful that I had the opportunity to attend that little school with them.

Lastly, I hope my readers will take note of their heritage and appreciate it as much as I do mine. I truly believe all things happen by design and that mostly we have a bit of a say in how we respond to our life's circumstances. Had I responded differently to my "unlit cigarette" situation by requesting a court martial, there is no doubt I would have reached the top of the enlisted ladder, and risen to the officer corps of the Navy's Medical Service Corps, because of the influence of my Daddy, Ben Morgan, Ms. Smith-Patton, Eddie and East Polkton High School.

"And the class of 57 had it dreams, we all thought we'd change the world with our great works and deeds, or maybe we just thought the world would change to fit our means, the class of 57 had it's dreams". (The Statler Brothers)

APPENDIX

List all areas where information is derived from:

Military Personnel records:
Address: National Personnel Records Center
1 Archive Drive
St. Louis, MO 63138-1002

- Anson County Record of Vital Statistics, Wadesboro, N.C.

- Union County Record of Vital Statistics, Monroe, North Carolina

- Mecklenburg County Record of Vital Statistics, Charlotte, North Carolina

- NC Department of Education, Raleigh, North Carolina

- USS Guam (LPH-9) Association

- Navy Cruise Books.com

- Recruit Training Command, Great Lakes, Illinois, Company 51 Year Book

- 1st Battalion, 2nd Marines LANFORMED 1962-63 Cruise Book

- USS Notable MSO 460 Deployment Logs 1967-68

- USS Ability MSO 519 Deployment Logs 1968-68

- USS Sanctuary AH 17 1969-70 Deployment logs

- USS Guam LPH 9 Special Deployment 1976-77 Cruise Book

Other works by DeeJak's Publishing Company:

I'm Finally A Man

A Husband's Journey
to Manhood

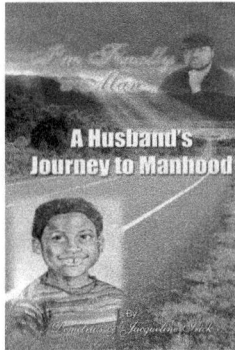

By Demetrius and Jacqueline Irick

This book is about the poor decisions made by a young adolescent and the consequences he faced as a result of his actions. The book also takes you through the challenges a young wife faced while trying to break the mindset and generational curses – in order to encourage an immature male into becoming a man.

I'm Finally A Man Prologue

I envisioned my body in a casket. My family, high-school classmates, and friends were packed in the old church. My skin was "black as soot" as the old people used to say. My once-caramel skin tone was now altered by the embalming fluid. I felt so terrible leaving this life behind; my parents, brothers, and friends looked to be taking it very hard.

The pain caused a burning sensation in my chest. There was a lump in my throat from seeing my mother's tears. The emotions I felt would not allow any words to come out as I tried to comfort her. As I scanned the church, I heard whispers of people talking about the suicide. "Why did he kill himself?" they asked.

"No!" I yelled.

Then I awoke suddenly from my sleep. Tears ran down my face. I was covered in sweat. The hair on my arms was standing straight up. "It's not fair; I'm too young to die!" I said before realizing it was just another nightmare. I felt like I was losing my mind. The signs had been there for months. I begged for help. I was slipping into a depression. I was in unfamiliar territory, struggling to distinguish reality from my nightmares. I had visions of the destruction of the world as we know it. I saw a world filled with death, violence, and nuclear war. All of humanity was struggling to live each day. I saw a world of good versus evil, angels and demons. It was the apocalypse. A spiritual war. I felt trapped, alone, and afraid in a world that was very unfamiliar.

As I interacted with family and friends, I felt like I was being treated as if I had the plague or three eyes. I felt abandoned by my family and friends. I prayed constantly, begging God to show me my purpose. I begged Him to take the stress away from me. I prayed that He would restore my mental competence. I prayed that He would give me a sign that He had plans for my life.

I grabbed my father's .32-caliber revolver from his dresser drawer. The house was so quiet it was eerie. I reflected on my latest dream and felt an inexplicable peace. I felt guilty for leaving my family and friends behind, but my mind was at peace. I couldn't remember the last night that I hadn't had a nightmare. I couldn't tell you the last time I felt in control of my thoughts. As I stared at the gun in my hand, I weighed my options: Should I continue down the path of nightmares, loneliness, betrayal, and feeling out of place? Should I seek the tranquil energy and peace I remembered feeling during my dream? I asked myself, How did I get here?

I had graduated high school, joined the military, and attended college. Now I was looking down the barrel of a gun. How is this God's plan? What does He want from me?

"I want my life back!" I yelled at this invisible God. "Why would you allow this to happen to me? What cruel Father would do this to someone He loved?" I felt the anger boiling in my blood as it coursed through my veins. I thought of the look of disgust I saw in the eyes of those who didn't understand what I was going through.

I made one last final appeal to this God, the God of my parents. I placed one bullet into the chamber of the revolver and spun the cylinder. I raised the gun to my head, closed my eyes and felt my heart pounding in my chest. "Dear God, if you have any plans for my life, please show me now. God, if you are real, do something supernatural. Make your presence known to me in plain English. Father, give me a sign even an idiot can understand." I pulled the hammer back on the revolver and prayed for forgiveness.

Live with Purpose
Making More Meaningful Decisions

By Jacqueline Irick

Over the years, as one matures and encounters different situations, he or she will have a better sense of purpose. Our purpose will cause us to become more selective about the circle of people with whom we associate. We will no longer have a use for certain behaviors and misdeeds in our life. However, many people may view this change as egotistical because they remain on the same beaten path and refuse to change. Moreover, this transition will result in us outgrowing some people with whom we were once close. Some family and friends will not understand this; however, everyone is not meant to accompany us through each journey in our life. Each journey will mark a new phase and once we've read and understood a chapter of our life, rarely do we go back.

DeeJak's Publishing
Social Media and Affiliated Businesses

follow us: @deejakspub
facebook.com/deejakspub
pinterest.com/deejakspub
www.deejakspub.wordpress.com

What is Coaching?

Coaching is a helpful way to develop yourself as a person, work through pivotal changes, or live a more focused life. I'm looking forward to working with you to help you cultivate a life of greatness!

Visit us at www.btccllc.com for more information

Jacqueline Irick – Event Planner and Caterer

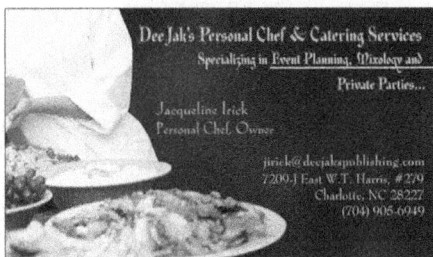

Workshop/ Webinar
Getting Started:
"How to Get Your Work Published"

Are you unclear about the publication process? Attend one of our workshops and get educated on writing your first book. We will cover the grant of rights, the author's rights and responsibilities, our expectations, how the author gets paid, formatting your material, getting your work registered and copy written & more. Workshops are held every month. Visit us at www.deejakspublishing.eventbrite.com for more information or www.deejakspublishing.com.

www.ingramcontent.com/pod-product-compliance
Lightning Source LLC
Chambersburg PA
CBHW021124020426
42331CB00005B/623